# Polluted Groundwater

# WATER INFORMATION CENTER, Inc.

## PERIODICALS

*Water Newsletter*
*Research and Development News*
*Ground Water Newsletter*

## BOOKS

Geraghty, Miller, van der Leeden, and Troise – *Water Atlas of the United States*
Todd – *The Water Encyclopedia*
van der Leeden – *Ground Water – A Selected Bibliography*
Giefer and Todd – *Water Publications of State Agencies*
Soil Conservation Service – *Drainage of Agricultural Land*
Gray – *Handbook on the Principles of Hydrology*
National Water Commission – *Water Policies for the Future*
Officials of NOAA – *Climates of the States*
The Water Research Association – *Groundwater Pollution in Europe*
Litton, Tetlow, Sorensen and Beatty – *Water and Landscape*
Meta Systems, Inc. – *Systems Analysis in Water Resources Planning*
Giefer – *Sources of Information in Water Resources*
van der Leeden – *Water Resources of the World*
Todd and McNulty – *Polluted Groundwater*

# Polluted Groundwater

## A Review of the Significant Literature

*David Keith Todd*

Department of Civil Engineering
University of California, Berkeley

*Daniel E. Orren McNulty*

Boalt Hall School of Law
University of California, Berkeley

Water Information Center, Inc.
Port Washington, New York

This book is based upon the text of report EPA-600/4-74-001, *Polluted Ground-water: A Review of the Significant Literature,* which was released by the U. S. Environmental Protection Agency in March 1974.

This printing has been undertaken to make this excellent report available in a conventional format for general and reference use.

Published 1976 by Water Information Center, Inc., 14 Vanderventer Ave., Port Washington, N.Y. 11050

Library of Congress Catalog Card Number: 75-21041
ISBN: 0-912394-16-1

# Preface

The current concern for control of water pollution and for the maintenance of high quality water supplies has stimulated an interest in the pollution of groundwater. This volume contains a selective review of the literature on man-caused groundwater pollution, including causes, occurrences, procedures for control, and methods for monitoring. We have made no attempt to develop a comprehensive bibliography on the subject; references were selected for inclusion on the basis of their significance and relevance.

Bibliographies and important general references are discussed separately. Thereafter the literature is described in essay form on a subject basis. References cited by number in the text are listed in complete bibliographic form at the end of the book together with an author index. With few exceptions, the material reviewed is limited to relatively recent published items in the United States. Administrative regulations, legal reports, and unpublished materials such as theses have been omitted.

The material in this book was originally prepared for the U. S. Environmental Protection Agency as part of a contractual study by TEMPO, the Center for Advanced Studies of the General Electric Company. Mr. Charles F. Meyer of General Electric-TEMPO and Mr. Leslie G. McMillion of the Environmental Protection Agency provided technical guidance during the course of the study. We are pleased to acknowledge their assistance as well as their encouragement to publish this work in book form. Finally, we wish to acknowledge the extensive use made of the libraries of the University of California, Berkeley, and particularly of the Water Resources Center Archives.

*David Keith Todd*
*Daniel E. Orren McNulty*

Berkeley
May 1975

# Contents

# CHAPTER I

# INTRODUCTION

This review is concerned with groundwater pollution resulting from activities of man. Natural groundwater, unaffected by man, contains salts in solution with the concentration depending upon the previous history of the water and upon geologic and hydrologic influences. Current governmental endeavors are focused on modifying the consequences of man's activities so as to prevent, reduce, and eliminate groundwater pollution, and to restore and maintain the integrity of the Nation's groundwaters.

Literature included herein was selected on the basis of its significance and relevance from a variety of bibliographies, general references, and abstracts. Where an abstract was not available to judge the value of the reference, the original reference was consulted to determine its pertinence. In some cases, only the title of a reference could be located in available libraries and within the time constraints of the study. Where the title appeared to so warrant, the reference was included at the end of the "References" list (numbers 596 to 661) but was not discussed in the text.

Excluded from consideration were all unpublished materials, administrative regulations at all governmental levels, and legal reports. The study has been limited to the literature of the United States, with the exception of a few Canadian reports and a recent book on the European groundwater pollution situation. Items of historical interest have also been excluded because they have limited bearing on the current and future pollution situation; only a few references predate 1950.

In the area of research reports, selections were restricted to those directly concerning changes in groundwater quality. Numerous works which indirectly relate to the subject, including flow and mixing phenomena in porous media, infiltration and clogging rates, adsorptive and ion exchange properties of soils, etc., were excluded.

Chapters II and III contain annotated listings of bibliographies and important general references. Thereafter, the literature is reviewed in essay form on a subject basis.

Seven subject headings are covered in Chapters IV to X. Under each of these, several subsections review literature on a given topic. Each of the 595 references reviewed is classified under one of the 30 topical headings. Many papers and reports embrace more than one subject; these have been assigned to the section which seemed most appropriate. Consequently, in using this review a reader is advised to consider closely related topics which may contain reviews that prove pertinent to his particular interest. Note also the "General" category in Chapter IX, which contains references on all facets of groundwater pollution.

At the end of the report all references cited by number are listed in complete bibliographic form, in the order mentioned in the text. Finally, an author index of all references is included to facilitate location of particular materials.

# CHAPTER II

# BIBLIOGRAPHIES

The following recently published bibliographies are closely related to the subject of groundwater pollution. Annotations describe the scope and extent of material included in each bibliography.

(A) Rima, D. R., E. B. Chase, and B. M. Myers. *Subsurface Waste Disposal by Means of Wells–A Selective Annotated Bibliography.* Water-Supply Paper 2020. U.S. Geological Survey, Washington, D. C., 1971, 305p.

A total of 692 references with abstracts covering source materials through 1969 are included. The references are about equally divided among three topics: disposal of oil-field brines, research on disposal of radioactive wastes, and case histories of industrial injection wells. Abstracts are arranged alphabetically by author. Geographic and subject indexes conclude the bibliography.

(B) *Subsurface Water Pollution–A Selective Annotated Bibliography,* Pt. I–Subsurface Waste Injection, Pt. II–Saline Water Intrusion, Pt. III–Percolation from Surface Sources. U.S. Environmental Protection Agency, Washington, D. C., March 1972, 156p., 161p., 162p.

These three volumes contain a total of 319 references; all are directly related to groundwater pollution. References include abstracts and are arranged according to WRSIC accession numbers. Each part of the bibliography includes a significant descriptor index and a comprehensive subject index.

(C) *Subsurface Water Pollution by Percolation–Selected Abstracts.* National Technical Information Service. Springfield, Virginia. Rept. NTIS-PK-134. November 1972. 35p.

This listing contains 35 references with abstracts and descriptors.

(D) Bader, J. S., and others. *Selected References–Ground-Water Contamination, The United States of America and Puerto Rico.* U.S. Geological Survey, Washington, D. C., 1973, 103p.

This contains 834 references without abstracts. Numerous cooperative area investigations by the U.S. Geological Survey are listed alphabetically by author. Particularly valuable are indexes according to geographic areas, states, kinds of contamination, sources of contamination, and general discussions.

(E) Summers, W. K., and Z. Spiegel. *Ground Water Pollution, A Bibliography*. Ann Arbor Science Publishers, Ann Arbor, Michigan, 1974, 83p.

More than 400 partially annotated references are listed in this bibliography. The material is organized on a subject basis under 12 chapter headings. The references concentrate on nitrates, heavy metals, and pesticides as well as on effects of urbanization, solid waste disposal, animal wastes, and petroleum products.

(F) WRSIC Bibliographies. In 1971 the Water Resources Scientific Information Center began publication of a series of bibliographies in water resources produced from the extensive information base comprising *Selected Water Resources Abstracts.*

Each bibliography includes a significant descriptor index, a comprehensive subject index, and an author index. References include abstracts and are arranged according to WRSIC accession numbers. The following bibliographies have been selected as being those which may or do contain references pertaining to groundwater pollution. The three bibliographies listed under the heading *Subsurface Water Pollution* are identical to those described in (B) above. For identification purposes the WRSIC numbers follow the titles. The PB numbers indicate availability from the National Technical Information Service.

| | | |
|---|---|---|
| *Strontium in Water* | WRSIC 71-201 | PB 201268 |
| *Arsenic and Lead in Water* | WRSIC 71-209 | PB 202578 |
| *DDT in Water* | WRSIC 71-211 | PB 212262 |
| *Detergents in Water* | WRSIC 71-214 | PB 206527 |
| *Dieldrin In Water* | WRSIC 72-202 | PB 207339 |
| *Aldrin and Endrin in Water* | WRSIC 72-203 | PB 210922 |
| *Chromium in Water* | WRSIC 72-205 | PB 210921 |
| *Mercury in Water* | WRSIC 72-207 | PB 206535 |
| *Soil Nitrogen Cycle* | WRSIC 72-208 | PB 209931 |
| *Sanitary Landfills* | WRSIC 72-214 | PB 211565 |
| *Subsurface Water Pollution* | | |
|   I.  *Subsurface Waste Injection* | WRSIC 72-220E | PB 211340 |
|  II.  *Saline Water Intrusion* | WRSIC 72-221E | PB 211341 |

| | | |
|---|---|---|
| III.  *Percolation from Surface Sources* | WRSIC 72-222E | PB 211342 |
| *PCB in Water* | WRSIC 73-201 | PB 217859 |
| *Artificial Recharge of Groundwater* | WRSIC 73-202 | PB 221479 |
| *Cadmium in Water* | WRSIC 73-209 | PB 218829 |
| *Water Reuse, Volume 1* | WRSIC 73-215 | PB 221998 |
| *Water Reuse, Volume 2* | WRSIC 73-215 | PB 221999 |
| *Phosphorus Removal, Volume 1* | WRSIC 73-208 | PB 221477 |
| *Phosphorus Removal, Volume 2* | WRSIC 73-208 | PB 221478 |
| *Acid Mine Water* | WRSIC 75-202 | |
| *Water Reuse, Volume 3* | WRSIC 75-204 | |
| *Water Reuse, Volume 4* | WRSIC 75-204 | |

# CHAPTER III

# IMPORTANT
# GENERAL REFERENCES

Among the numerous references relating to groundwater pollution, a few view the subject broadly and comprehensively. Because such general references are important for persons wishing to obtain an introduction to the subject, these publications have been specially listed and described below.

(A) *Ground Water Contamination, Proceedings of the 1961 Symposium.* Public Health Service, U.S. Department of Health, Education, and Welfare. Tech. Rept. W61-5. 1961. 218 p.

This report summarizes papers and discussions of a Public Health Service symposium on groundwater contamination. A total of 38 papers were presented, organized around five topics: (1) Hydrogeological aspects of groundwater contamination, (2) Types of contaminants, (3) Specific incidents of contaminants in groundwater, (4) Regulations and their administration, and (5) Research on groundwater contamination. Some 192 references are included. This is the earliest comprehensive analysis of the many facets of groundwater pollution.

(B) Ballentine, R. K., S. R. Reznek, and C. W. Hall. *Subsurface Pollution Problems in the United States.* U.S. Environmental Protection Agency. Washington, D. C. Tech Studies Rept. TS-00-72-02. May 1972. 29 p.

A general discussion of the subject is presented under the headings of deep well injection, percolation from surface sources, salt water intrusion, and controls.

(C) Fuhriman, D. K., and J. R. Barton. *Ground Water Pollution in Arizona, California, Nevada, and Utah.* Fuhriman, Barton and Assocs. Washington, D. C. Water Pollution Research Series Rept. 16060 ERU 12/71. U.S. Environmental Protection Agency. December 1971. 249 p.

This report covers groundwater pollution problems in Arizona, California, Nevada, and Utah. Natural mineralization is mentioned as the most important influence on groundwater quality. Significant man-caused effects include irrigation return flows, sea water intrusion, solid wastes, and disposal of oil field

6

brines. Research needs to control pollution are discussed. The report includes a list of 241 references, a glossary of terms, a summary of water quality standards for various water uses, and a bibliography of 1132 items.

(D)  Scalf, M. R., J. W. Keeley, and C. J. LaFevers. *Ground Water Pollution in the South Central States.* U.S. Environment Protection Agency. Corvallis, Oregon. Rept. EPA-R2-73-268. June 1973. 181 p.

This report describes present and potential groundwater pollution problems of Arkansas, Louisiana, New Mexico, Oklahoma, and Texas. Mineralization due to natural causes is listed as the most influential factor on groundwater quality, while oil field activities constitute the greatest man-made cause. Research needs to solve problems associated with various causes of underground pollution are enumerated. The report includes a list of 132 references, a glossary of terms, a summary of water quality standards for various water uses, and a bibliography of 385 items.

(E)  *Ground-Water Contamination–An Explanation of Its Causes and Effects.* Geraghty & Miller, Inc. Port Washington, N.Y. May 1972. 15 p.

This pamphlet gives a general description of groundwater pollution. Included are causes and types of contamination, geologic influences and the movement of pollutants, governmental regulation, and investigations for control. A suggested reading list of 33 items concludes the report.

(F)  Pettyjohn, W. A. (editor). *Water Quality in a Stressed Environment.* Minneapolis, Minn., Burgess, 1973. 309 p.

This book consists of a collection of previously published papers on water pollution. Geologic controls associated with groundwater pollution are the subject of five papers, while seven others describe examples of groundwater pollution. Included in the examples are reports on pollution from industrial plants, chemical plants, oil field brines, sewage lagoons, horse stables, septic tanks, and bacteria and viruses.

(G)  Campbell, M. D., and J. H. Lehr. Ground Water Pollution. In: *Water Well Technology.* New York, N.Y., McGraw-Hill, 1973. p. 11-28.

This chapter gives a brief review of the causes of groundwater pollution with emphasis on effects of wells. A list of 32 references is included.

(H) Meyer, C. F. (editor). *Polluted Groundwater: Some Causes, Effects, Controls, and Monitoring.* General Electric Company. Santa Barbara, Calif. Rept. EPA-600/4-73-001b. U.S. Environmental Protection Agency. July 1973. 264 p.

This comprehensive report describes methods for controlling groundwater pollution resulting from injection wells into saline water and freshwater aquifers; land disposal and septic systems; sewer, tank, and pipeline leakage; surface waters, the atmosphere, and urban areas; salt water intrusion in coastal and inland aquifers; and spills and artificial recharge. In addition, for each causal factor the environmental consequences, pollution movement, and monitoring procedures are discussed. The report includes 33 figures, 29 tables, and 256 references.

(I) WPCF Research Committee. 1972 Water Pollution Control Literature Review–Effects on Groundwater. *Jour. Water Pollution Control Federation. 45:* 1296-1301, June 1973.

This annual review (which appears in each June issue) summarizes briefly the recent groundwater pollution literature. A total of 30 references are discussed, almost all having appeared in 1972. (It should be noted that other sections of the review may also contain pertinent material: continuous monitoring, automated analysis, and sampling procedures; lagoons and oxidation ponds; detergents; water reclamation and reuse; deep-well injection; agricultural wastes; solid wastes and water quality; and radioactive wastes.)

(J) Miller, D. W., F. A. DeLuca, and T. L. Tessier. *Ground Water Contamination in the Northeast States.* U.S. Environmental Protection Agency. Ada, Oklahoma. Rept. EPA-660/2-74-056. June 1974. 325 p.

This report evaluates the principal sources of groundwater pollution in the 11 northeast states, including all of New England, New York, New Jersey, Pennsylvania, Maryland, and Delaware. Principal sources of quality degradation include septic tanks, buried tanks and pipelines including sewers, highway deicing salts, and landfills. The report includes a list of 403 references, a glossary of terms, and a summary of drinking water standards.

(K) Cole, J. A. (editor). *Groundwater Pollution in Europe.* Proceedings of a Conference organized by the Water Research Association in Reading, England, September 1972. Port Washington, N. Y., Water Information Center, 1974. 547 p.

This comprehensive volume contains some 52 papers and case histories concerning all aspects of groundwater pollution in Europe. The subject matter is organized under the following topical headings: legal and administrative measures, hydrogeology and hydraulics, chemistry, microbiology, case histories, tracers, and deep well disposal. An extensive bibliography and a subject index complete the book.

# CHAPTER IV

# URBAN POLLUTION

## EFFLUENT RECHARGE

Effluent from municipal wastewater treatment plants often is discharged into surface waters. However, in some instances the treated water is reclaimed by percolating it into the ground to recharge aquifers. The groundwater pollution possibilities inherent in water reclamation by artificial recharge projects have been explored frequently.[1-9] In 1955 the University of California Sanitary Engineering Research Laboratory[1] gathered and evaluated pertinent studies, and reviewed methods and statistics of recharge by effluent spreading and injection. Data on infiltration rates and pollution travel were cited, along with the engineering and economic aspects involved.

In 1968 Popkin and Bendixen[2] summarized studies on the application of liquid waste to the soil, in which continuing hydraulic acceptance and percolate quality were stressed. Results suggested that design and operation of a soil adsorption system could be improved by weekly dosing and/or by use of improved pretreatment processes.

Tchobanoglous and Eliassen[3] in 1969 discussed various factors related to the indirect cycle of water reuse. Methods of treated wastewater recharge were surface spreading, direct injection, and pits and leach field seepage. Recharge operations required consideration of the rate, quality, and quantity of wastewater application; site characteristics; and available treatment processes. Finally, a cost-benefit analysis for economic feasibility of indirect reuse of reclaimed water was outlined.

In 1969 Bouwer[4] described how aerobic percolation and subsequent lateral movement of low quality water could remove biodegradable materials, pathogenic organisms, and certain inorganic substances. With respect to problems of recharge basin management, the report warned against allowing accumulation of suspended materials on the basin bottom and against insufficient oxygen reaching the soil during dryups. Basins also needed to avoid excessive water table buildup while achieving maximum recharge per unit area. Nitrate reduction by denitrification was identified as a major problem in renovating sewage effluent.

The geohydrology of liquid waste disposal by irrigation was reviewed by Born and Stephenson[5] in 1969. The thickness, nature, and distribution of unconsolidated surface deposits determined infiltration, adsorption storage, and downward movement of wastewater. The uses of infiltrometer tests, laboratory examinations, and flow systems were explained as methods of monitoring wastewater recharge.

Additional monitoring and control methods were analyzed by Martin[6] in 1969. The importance of soil surveys in minimizing leaching, erosion, and groundwater contamination was stressed, and a Minnesota soil survey was included. Nitrogen contamination problems could be controlled by anaerobic conditions, plant growth, holding lagoons, and rotation spreading. In addition, several successful land waste disposal systems were described.

Dvoracek and Wheaton[7] in 1970 presented various localized examples of groundwater contamination by artificial recharge due to poor quality recharge water. Various methods of recharge were described (including wells, shafts, holes, pits, trenches, spreading, and "clean" nuclear explosions), and the contamination potential of each was discussed.

More recently, in 1972 and 1973, planning and design criteria for specific waste disposal methods have been presented. The Pennsylvania Bureau of Water Quality Management[8] published a manual on spray irrigation methods and designs. The design and engineering aspects of a proposed operation, site selection criteria, and essential groundwater quality monitoring data were all detailed.

Similarly, Bernhart[9] analyzed various soil infiltration and evapotranspiration methods of wastewater disposal. Project design and area calculations for seepage beds were included, and the effectiveness of septic tanks, aeration tanks, conventional tile fields, and seepage beds were charted. The study also considered the "horizontal protective distance" required for water supply wells under various conditions.

More localized studies of effluent recharge and disposal problems and practices have also been done in many states since 1950. Two Pennsylvania State University studies in 1967-68 dealt with the renovation of wastewater effluent by irrigation of forest land. Pennypacker, et al.[10] conducted a field study of treated sewage effluent sprayed on forest land, and found that while ABS and phosphorus were removed in the top soil layers, greater depths were required to remove nitrate, potassium, calcium, magnesium, and sodium. Sopper[11] applied treated municipal wastewater to forested areas, and achieved satisfactory renovation at rates up to four inches per week during April-November. MBAS concentration in the effluent exceeded standards for potable water, but was decreased to safe levels after passing through the forest floor and six inches of mineral soil. Approximately 90 percent of the applied wastewater was recharged to the groundwater reservoir at an application rate of two inches per week.

A review of the sewage disposal system of St. Charles, Charles County, Maryland, appeared in *Ground Water Age*[12] in 1973. Wastewater had been renovated to approximate drinking water by means of sewage lagoon systems supplemented by spray irrigation procedures.

In 1968 Bendixen, et al.[13] reported on the monitoring of a municipal ridge and furrow liquid waste disposal system in Westby, Wisconsin. Four one-acre basins disposed trickling filter effluent into the soil, and a heavy stand of unharvested grass apparently contributed to successful operation. The changes in infiltration rates and infiltrate quality due to season and various loading and operating conditions were examined.

Ketelle[14] in 1971 presented a general discussion of hydrologic and geologic factors relating to liquid waste disposal, with a case study in Southeastern Wisconsin. The geography, climate, geology, soils, and groundwater of a seven-county region were analyzed, and a final map of the area was developed indicating suitability of areas for liquid waste disposal.

Muskegon County, Michigan, was the site of research preparations by Chaiken, et al.[15] in 1972 to study the land disposal of treated sewage by spraying. A detailed observation well network was set up consisting of about 300 wells along the storage lagoon ditches and the site periphery. Samples were to be analyzed for 43 physical, biological, and chemical water quality parameters.

In 1968 Harvey and Skelton[16] summarized a field study at a Springfield, Missouri, sewage plant. Secondary treated effluent traveled underground where aeration was impossible. The turbidity and odor of the groundwater made it unsuitable for most purposes. Seepage runs, dye studies, and a seismograph survey were used to determine the source of the pollution.

Brown and Signor[17] in 1972 surveyed the principles and methods of groundwater recharge in the Southern High Plains of New Mexico and Texas. Artificial recharge pollution hazards included recharge waters with high particulate matter concentrations and faulty design and construction of wells.

An "overland-flow" sprinkler irrigation system successfully disposed of the wastewaters of a cannery plant in Paris, Texas. A 1973 article in *Water and Sewage Works*[18] reported a 99 percent reduction in BOD and up to 90 percent reduction in nitrogen from vegetable and grease wastes. The 640 acres were designed to accept an application rate of one-quarter to one-half inch per day. No underground migration of pollutants occurred due to downslope percolation, terrace collection, and channels to a receiving stream.

An experimental project on reclaiming water from secondary sewage effluent with infiltration basins in the dry Salt River bed near Phoenix, Arizona, has been uniquely successful (the Flushing Meadows Project). Bouwer[19, 20, 21, 22] and Bouwer, et al.[23, 24] have described this project from 1968 to 1972. The hydrogeology of the Salt River bed was very suitable for high rate wastewater reclamation by groundwater recharge. The project contained six recharge basins, with infiltration rates decreasing from grass to gravel to bare soil basins. Effective removal of 90 percent of nitrogen was obtained with long inundation periods, and the usual reductions in BOD, coliforms, and phosphorus were observed. In addition, these studies indicated that the cost of surface-spreading renovation of wastewater for groundwater recharge was comparable to tertiary in-plant treatment costs.

Hydraulic properties of the Salt River aquifer (including anisotropy) were evaluated by electric analog, and plans for a large scale project consisting of central collection wells flanked by strips of recharge basins on both sides of the river bed were developed. Bouwer[21] in 1970 presented criteria for design of such a system: (1) a maximum limit for the elevation of the water table mound beneath the spreading areas; (2) a minimum limit for underground detention time and travel distance to the wells; and (3) a minimum contamination of the groundwater in the aquifer outside the recharge system.

In 1968 Wilson, et al.[25] examined the dilution of an industrial waste effluent with river water in a thick vadose alluvium during pit recharge near the Santa Cruz River, Arizona. Water content profiles and groundwater hydrographs were observed near the recharge site and an abutting ephemeral stream. The report concluded that recharging of highly concentrated waters should coincide with periods of fully developed river recharge mounds and/or when ephemeral streams were discharging, and recommended evaluation

of pumping as a mixing and blending procedure. The significance
of groundwater quality monitoring was also emphasized.

Sewage reclamation projects by artificial recharge throughout
California have been extensively documented since 1950. Stone
and Barber[26] described the infiltration of sewage effluent through
a spreading basin in Los Angeles County in 1951. It was found
that if aerobic conditions were maintained in the percolating fluid,
successive increases in TDS were small enough to allow the wastewater to be reused two to five times. Standards for dissolved oxygen and BOD contents in the percolated fluids to insure satisfactory
groundwater were also discussed.

In 1953, the California State Water Pollution Control Board[27]
reported on bacterial and chemical pollution at a wastewater reclamation project at Lodi. Bacteriologically safe water from settled
sewage or final effluent resulted from passage through at least four
feet of soil, and water of satisfactory chemical quality was achieved
when the raw sewage did not contain high concentrations of undesirable industrial wastes. Maximum safe percolation rates, various
spreading techniques, and side effects (e.g. mosquitoes, algae, and
odors) of wastewater reclamation were also discussed.

Stone[28] on 1953 surveyed methods of land disposal of domestic sewage and industrial wastes throughout California. The need
for an aerobic environment was stressed in every method, and stand-
by lagoons, spreading basins, and irrigation fields were recommended for peak and emergency effluent loads. No cases of groundwater contamination were reported in 69 communities having sewage farms.

In 1958 the California Department of Water Resources[29] presented data on sewage treatment facilities and the status of existing
and proposed reclamation projects. Examples included proposed
spreading basins to recharge the San Luis Ray groundwater basin.

McMichael and McKee[30] conducted a 1962-65 field investigation and laboratory study of percolation of municipal wastewater
effluents at Whittier Narrows, Los Angeles County, California.
Chemical analyses of 25 test wells monitored at various depths and
of a sampling pans beneath the basins revealed satisfactory degradation of ABS under aerobic conditions, while nitrates and chlorides
met USPHS standards. The successful control methods consisted
of activated sludge plants combined with a six-inch layer of pea
gravel on the spreading basins.

In 1966 Doneen[31] summarized a field investigation of the
native salts in the substrata of the west side and trough of the San

Joaquin Valley of California. A proposed program of cyclic use and storage of groundwater involving recharge of underground storage basins in the area was shown to be extremely hazardous. A summary of data at three typical sites disclosed that gypsum, salines, and exchangeable sodium in the substrata would cause percolating or recharge waters reaching the groundwater to be of very poor quality.

In 1971 Matthews and Franks[32] reported on tests at the "Cinder Cone" sewage disposal area at North Lake Tahoe, California. Geologic and hydrologic features of the region were evaluated, and results of drill hole testing indicated that the subsurface would accept the sewage effluent. Trenches were made in the area, and percolation rates were studied to discover effective filtration and treatment. In general, no change in groundwater quality was detected in the sampling areas.

Boen, et al.[33, 34, 35] in 1971 completed a project which investigated the feasibility and safety of neutralizing wastewater recycled through the groundwater of the Hemet-San Jacinto Valley of California. An additional objective of the six-year study was to analyze any benefits to the area salt balance problem caused by recharging. Yearly quantities of wastewater reclaimed were given, and the lack of groundwater pollution at surrounding water wells was attributed to the inhomogeneous nature of the basin geology. The study added invaluable knowledge to the technology of intermittent wastewater percolation and associated monitoring techniques. In addition, a novel feature of the project was the employment of highly sensitive temperature probes to trace the lateral migration of the recharged water.

In 1972 Young, et al.[36] outlined a planned study of wastewater reclamation by irrigation on Oahu, Hawaii. Wastewater recharge by sprinklers was contemplated, and the effects of virus and salt movement on groundwater quality were to be analyzed.

## LANDFILLS

Municipal dumps and sanitary landfills have long been recognized as potential sources of groundwater pollution; however, little quantitative information as to their specific effects was available. To rectify this situation a series of studies was undertaken in California under sponsorship of various state agencies. The earliest of these was a study of the sanitary landfill at Riverside, California, in 1953-54.[37, 38] Field measurements showed that pollution was limited to small increases in total dissolved solids where the water table was in contact with the landfill. Rainfall at that location was not sufficient to produce leachate. Pollution moved in the direction of groundwater flow, showed limited vertical mixing, and was confined to the shallowest portion of the aquifer. Pollution was detected as far as one-half mile downstream of the landfill. Most gas formed in the landfill escaped to the atmosphere.

In a California review of groundwater pollution from refuse dumps in 1961,[39] the causes of pollution were identified as infiltration and percolation of surface water, refuse decomposition, gas production and movement, leaching, and groundwater movement. Research projects were recommended to provide information so that pollution could be minimized.

A field investigation of the production of gases in a landfill and its effect on groundwater quality at Azusa, California, was reported in 1965.[40] Sizable $CO_2$ concentrations can be expected in the bottom layer of refuse for many years. The concentration of $CO_2$ in groundwater depended upon the depth to the water table and the groundwater flow rate. Forms of gas control considered were liners, fill, soil injection, ventilation, and burning.

In 1969 the California Department of Water Resources reported on four experimental landfills in California,[41] while Coe summarized the results in 1970.[42] Groundwater impairment was typified by temporary increases in organic material and permanent increases in total dissolved solids, chloride, sulfate, and, in addition, hardness and bicarbonate from effects of $CO_2$. It was recommended that sanitary landfills should be designed as a system with primary concern given to site selection, materials to be deposited, construction and operation techniques, and use of the completed fill. A classification scheme was described for physical characteristics of a site according to the degree of protection afforded receiving waters and to the type of refuse to be disposed.

A detailed study of the hydrogeologic aspects of solid waste disposal sites was conducted in Northeastern Illinois by the Illinois State Geological Survey during the period 1967-70. Results of the study have led to a series of reports on hydrogeology of refuse sites, groundwater pollution, site selection, and design criteria. The initial report[43] described geologic environments in Northeastern Illinois and their relation to those considered safe for refuse disposal, namely, low permeability, relatively dry areas, and hydrologically protected sites.

Hughes, et al.[44] evaluated the hydrogeologic environments in the vicinity of four existing landfill sites in the Chicago area. Movement and dissolved solid contents of groundwater were determined; information obtained can be used by regulatory agencies to define suitable landfill sites. It was emphasized that groundwater flow systems must be determined if movement of refuse leachate is to be predicted and that this may be difficult except in homogeneous materials.

The problem of leachate pollution either from rainfall or high water tables was discussed based on the Illinois study by Landon in 1969.[45] He mentioned that liners for landfills are often impractical, costly, and can lead to problems when leaks develop. Preferred control alternatives are based on site selection and include: (a) knowledge of existing hydrogeologic conditions which would favorably control rate and direction of leachate migration, (b) engineering the landfill to collect and treat leachate, and (c) construction of limited collection facilities to supplement natural conditions.

A comprehensive report on the Illinois study was prepared by Hughes, et al.[46] in 1971. The distribution and concentration of dissolved solids in the vicinity of four landfills in Northeastern Illinois were measured and found to be controlled by the groundwater flow system. Attenuation of dissolved solids in groundwater after leaving the landfill was primarily influenced by the particle size of earth materials and the distance traveled. Precipitation was adequate to leach a completed landfill. It was concluded that where the natural environment is not capable of containing or assimilating leachate, a landfill can be made safe by lining the disposal site, by collecting and treating the leachate, or by other relatively simple engineering procedures.

A summary report of the same study[47] pointed out that 90 percent of Illinois is suitable for sanitary landfills because of fine-textured surficial materials and favorable locations within hydrogeologic flow systems; however, the remaining 10 percent contains most of the proposed landfill sites. It was stressed that although technology is available to handle solid waste disposal problems, regulations and their implementation are major needed requirements.

A by-product of the Illinois landfill study was a report by Cartwright and McComas in 1968[48] on the use of electrical resistivity and soil temperature surveys to measure groundwater pollution from landfill leachates. Comparisons were made with groundwater quality measurements in monitor wells. One resistivity survey traced mineralized water from a landfill for a distance of more than 1,000 feet and flow patterns agreed with interpretations based on monitor well data. The geophysical surveys showed, in general, that chemically altered water is traceable in uniform earth materials where the depth of the water table is constant. The soil temperature survey indicated the presence of a halo of higher temperatures around the landfill as well as areas of surface recharge.

The Illinois study produced detailed site evaluation criteria for landfills.[49] To protect groundwater and surface water, landfills should be located in relatively impermeable material to retard leachate movement, and there should be at least 30 feet of impermeable material between the bottom of the landfill and the shallowest aquifer. Proper site topography is important to avoid surface drainage contamination. Limestone quarries and sand and gravel pits make poor sites, as do poorly drained swampy areas. Strip mines, clay pits, and gravel pits with a high percentage of natural clay binder do make good disposal sites if kept dry. Flat upland areas are also good sites if a clay barrier is present above any aquifer. It was concluded that careful selection of a landfill site will result in little, if any, danger of groundwater pollution.

In 1969 Farvolden and Hughes,[50] using the Illinois study data, suggested sanitary landfill design criteria to minimize pollution of groundwater. Most important is to keep the landfill unsaturated. If this is not possible, hydrologic conditions must prevent fast migration of leachate or provide convergent flow toward collection sites; these could be natural slopes or ditches, drains, and pumping wells. Where refuse is piled to form a hill (for subsequent recreational use) groundwater flow in the vicinity is controlled by the groundwater mound that develops under the hill of refuse. Springs of objectionable leachate should be anticipated where the hill method is employed.

A comprehensive study of the effects of a landfill on groundwater quality was conducted at Brookings, South Dakota, during the period 1964-72 by Andersen and Dornbush.[51,52,53] Initial results led to conclusions that chloride, sodium, and specific conductance were the most useful parameters for detecting contamination and that because ionic concentrations increased during rainy seasons, effects of leaching overrode those of dilution. Water quality improved with distance downstream from the landfill; also, a trench constructed to intercept groundwater as it moved from the fill area acted to improve the water quality. Later results led to recommendations that disposal sites within the influence of pumping wells should be avoided, the deposition of refuse into ponds and at depths touching the water table should be avoided, that burning be minimized because it increases the permeability and hence the leaching of wastes, and that tight cover soils and good drainage should be provided to reduce leaching.

A survey of information on landfill pollution by Weaver in 1964[54] led to statements that leaching of refuse can produce organic, mineral, and bacteriological pollution, and that where refuse is in contact with a water table, the water may become unfit for domestic or irrigation use. Although bacterial and organic pollution may be limited in extent, chemical pollution—including methane, $CO_2$, ammonia, and hydrogen sulfide—may range over long distances.

In 1968 Lane and Parizek[55] reported on a detailed field investigation of a landfill near State College, Pennsylvania. The landfill is situated on steep slopes over a dry valley bottom with the water table about 250 feet deep. To monitor water quality, leachate was intercepted by a plastic sheet and carried to an infiltration trench. Suction lysimeters were installed at various depths in the soil beneath the landfill trench. Movement of a wave front of leachate-polluted soil water could be traced in the soil, indicating that severe pollution of soil water in the immediate vicinity of a landfill can result even though the landfill is not in direct contact with a water table and even before the refuse has become saturated to field capacity.

A later field study of landfill leachate at the same State College, Pennsylvania, site was reported by Apgar and Langmuir in 1971.[56] Samples of the leachate in the unsaturated sandy-clay to sandy-loam soils beneath the landfill were collected. The quantity and quality of leachate varied considerably with the topographic setting of landfill trenches. High values of specific conductance, chloride, BOD, nitrate-nitrogen, and iron were reported. Leachate moved downward at the rate of 6-11 ft/yr and was found to be highly contaminated at depths of 50 feet or more.

Remson, et al.[57] in 1968 analyzed the water movement in an unsaturated sanitary landfill. Moisture-routing methods were applied to predict vertical movement of moisture through a hypothetical landfill based upon climatological techniques and hydraulic properties of the fill and the overlying soil cover. Results showed that the time elapsed before the appearance of leachate depended on the season of emplacement and the initial moisture content.

A short review paper by Dair[58] in 1969 pointed out that although landfills in Southern California have been intensively investigated, problems of discovering feasible yet sanitary methods of depositing refuse in direct contact with groundwater, of evaluating the amount of leachate to water tables, and of developing barriers to prevent the escape of refuse-produced gases in groundwater still remain to be solved.

Qasim and Burchinal[59] in 1970 reported on analyses of leachate from simulated landfills consisting of 3-ft. diameter cylinders filled with mixed refuse, saturated with water, and with additional water added at two-week intervals. The concentration of leachate increased initially, began to decrease after four weeks, and increased again after eight weeks. Reports were presented on 18 organic and inorganic compounds over the 163 days of the study. Leach samples tended to undergo bacteriological and chemical self-purification.

Two existing sanitary landfill sites in Madison, Wisconsin, were examined by Kaufman[60] in 1970. He found that groundwater adjacent to the landfills received pollutants although adverse effects were limited. Groundwater recharge was between 35 to 50 percent of annual precipitation with lateral discharge to adjacent groundwater and surface-water resources. The increase in dissolved solids was high but restricted to local areas.

The basic pollution problems of solid waste disposal were reviewed in 1970 by Schneider.[61] The 1400 million pounds of solid wastes produced each day in the United States are disposed of by one of four methods: open dumps, sanitary landfill, incineration, and onsite disposal. Each method carries an inherent potential for water pollution. Seepage of rainwater through wastes leaches undesirable constituents which may then cause biological and chemical pollution of groundwater. Pollution potential is highest in permeable areas with a shallow water table where wastes are in direct contact with groundwater. Site selection for solid waste disposal must be based on adequate water resources information if pollution is to be minimized.

A similar survey of the hydrogeological aspects of selecting refuse disposal sites in Idaho was made by Williams and Wallace[62] in 1970. They recommended environments with low permeabilities, deep water tables, and protective engineered sites such as impermeable liners and covers.

A comprehensive review of groundwater pollution due to municipal dumps was prepared in 1971 by Hughes, et al.[63] Topics covered included groundwater pollution by solid wastes; significant research in this field; regulations; criteria for site selection; safeguards; and observation, detection, and identification of pollutants. Two bibliographies of over 600 references completed the report.

It should also be noted that a bibliography on sanitary landfill leachate travel was prepared by Emery[64] in 1971.

Design procedures for controlling groundwater pollution from sanitary landfills were described by Salvato, et al.[65] in 1971. Primary pollutants are BOD, COD, iron, chloride, and nitrate. Leachate originates as groundwater, surface water drainage, or precipitation; moisture within the refuse itself is only rarely adequate to produce movement. Detailed descriptions of means to control leachate from each source were presented, including impermeable barriers, surface and subsurface drains, and sumps with pumps. Diagrams illustrated the procedures.

Technical and economic aspects of community disposal systems and their effects on the environment were described by Sheffer, et al.[66] in 1971. Seven landfills in the United States were reviewed. A landfill stabilization project at Santa Clara, California, showed that aeration of sanitary landfills eliminated vermin and bacteria by high-temperature oxidation.

Fungaroli[67, 68, 69] developed a laboratory and a field sanitary landfill to provide information on the behavior of sanitary landfills in an environment common to the northeastern states. The long-range objectives of the study were: (1) to provide means for predicting movement of pollutants in subsurface regions under sanitary landfill sites; (2) to develop hydrologic, geologic, and soil criteria for the evaluation of site suitability for sanitary landfill operations; and (3) to appraise design methods and remedial procedures for reducing any undesirable contaminant movement. The reports described experimental facilities and contained experimental data.

A comprehensive and critical review of the important literature on the pollution potentials of groundwater from sanitary landfills and dump grounds was prepared by Zanoni[70, 71] in 1971. He reported that landfill leachate has highly pollutional characteristics; however, once in the underground the attenuating mechanisms of dilution, adsorption, and microbial degradation tend to reduce the impact on groundwater. Landfill practices in 21 states of the United States were described. A series of recommendations was presented for regulatory agencies concerned with approving and licensing solid waste disposal sites. Site selection, disposal, and construction procedures were included. It was emphasized that an agency should have a geologist to assist in site selection processes. Extreme caution should be exercised before approving ground disposal of industrial wastes. Monitor wells should be used where doubt exists as to the future effects of a landfill on groundwater. An agency should endeavor to minimize water percolation through refuse, thus encouraging leachate attenuation. The use of rock, gravel, or sand quarries for refuse disposal should be prohibited.

Field observations on a landfill at Moscow, Idaho, in 1970 by Seitz[72] provided information on groundwater pollution effects. It was found that groundwater in direct contact with solid waste exhibits dramatic increases in dissolved ions. But downgradient from the landfill, the leachate caused only a doubling of natural ionic concentrations and thus was still well below critical contamination levels. The reduction downgradient was attributed to low aquifer transmissivities, soil filtering, cation exchange, dilution, and utilization of ions by plants and soil bacteria. In the area studied leachate production and movement could be minimized if weathered granite could be avoided for landfill trenches, and if not, impermeable linings should be installed to retard water movement. Besides sampling from piezometers, the electrical resistivity method was tested and found satisfactory, provided traverses were nearly horizontal.

In a study of geohydrologic environments for solid waste disposal in Maryland, Otton[73] in 1972 reported on landfill leachate analyses from other states. The applicability of the five types of terrane in Maryland to solid waste disposal were described, including their soil and hydrologic characteristics. Suggested criteria for groundwater protection from landfill leachate included classifying sites and correlating them with the types of waste allowed. Landfill design, such as providing relatively impervious cover material, was also stressed. It was recommended that chemical and bacterial quality monitoring of groundwater be undertaken at three selected sites.

The most recent thinking on the effects of sanitary landfills on groundwater quality was contained in a summary of an engineering conference on the subject held in 1972.[74] Reviewing all available information, it was the opinion of participants that landfill leachate can be controlled and need not cause groundwater pollution. To achieve this goal, however, requires a properly engineered sanitary landfill involving site selection, cover material, and surface and subsurface leachate collection systems.

## ROAD SALTS

In 1970 Hanes, et al.[75] presented a literature review on the effects of salts and deicing salt additives on groundwater quality. Roadside groundwater and shallow well water samples revealed the effects of high chloride contents on plants and animals. The human health hazards still appeared speculative. Methods of control discussed included diversion ditches, better landscaping, and limited use of salts.

Surveys of the chloride in groundwaters of the Northeast due to the application of salts to highways have been done by Walker[76] and Struzeski.[77] Both included examples of threats and damages to domestic and industrial water supplies. In addition, Walker[76] suggested the inspection of salt storage sites and more efficient salt spreading equipment and procedures as means to control the problem.

In 1973 Field, et al.[78] surveyed groundwater pollution due to road salting in the Northeast and discussed various alternatives to present chemical melting procedures. These included: "snow melters", compressed air or high speed fluid streams in conjunction with snowplow blades or sweepers, snow/ice adhesion reducing (hydrophobic/icephobic) substances, and improved vehicular and/or tire design. Moreover, various types of salt storage facilities designed to reduce pollution were detailed and diagrammed.

Since 1970 numerous field studies of specific road salt pollution problems have been reported. Walker[79] detailed a 1955-70 investigation of chloride in the groundwater from pumping wells in Peoria, Illinois. The source of pollution was found to be a city street salt storage facility; control required containment of the salt and pumping of the chloride-affected groundwater to waste.

A 1965-69 study by Hutchinson[80] attempted to determine the environmental pollution resulting from an average annual application of 25 tons of sodium chloride to each mile of paved highway in Maine. Analyses of groundwater samples indicated that wells and farm ponds were seriously contaminated with chloride ions, while soil sample analyses revealed that soils contiguous to highways contained sodium levels that threatened vegetation and soil drainage.

In 1971 Broecker, et al.[81] summarized measurements of chloride ion content of ground and surface waters in the suburban area northwest of New York City. The purpose was to determine whether the temporal and geographic distribution of the chloride ion resulting from application of road deicing salts could be used to determine the time constant for groundwater renewal. The preliminary study clearly showed that chloride ion added to the groundwaters

of the New York area provided a valuable indicator of the retention time of soluble pollutants in the groundwater. Based on the
tonnage of salt applied and the percent of precipitation appearing
as runoff, the average chloride increase would be 40 ppm. Observations at one reservoir in the area showed a value of 30 ppm.

Chloride groundwater pollution by road salts in Massachusetts
was reviewed by Coogan[82] and Huling and Hollocher.[83] Coogan[82]
surveyed the problem since 1940 and concluded the sources of
contamination were salt storage piles and road drainage. He recommended storage of salts in buildings, not piles, and away from
groundwater supplies. Huling and Hollocher[83] sampled existing
wells in the suburban area of Boston, and found chloride contents
up to 100 mg/1, with higher values expected in wells near roads.
A gradual increase up to 1970 was observed, and the groundwater
had exceeded salt limits for persons on low-sodium diets.

In 1973 Dennis[84] reported on a 1967-73 examination of high
chlorinity groundwater in shallow wells of Indianapolis, Indiana.
Runoff from large quantities of deicing salts in 1966 was the cause
of pollution. Removal of the salt piles in 1968 was expected to
restore the groundwater to acceptable concentration levels by 1974.

## SEPTIC TANKS

The problem of groundwater contamination in unsewered
areas of Minnesota from 1950-1959 was attributed by Woodward,
et al.[85] to the widespread use of individual water supplies and sewerage systems (septic tanks and seepage pits). By 1959, over 40
percent of the water supplies of one Minneapolis suburb yielded
groundwater of high chemical content. The report included results
of a 1959 field study of 98 individual water supplies in the city of
Coon Rapids and 63,000 wells in the metropolitan Minneapolis area.
Water supplies were analyzed for nitrate, surfactant, coliform, and
chloride contents, and 47 percent of the 63,000 wells were found
to be contaminated. Proposed control methods included regulation
of individual construction, installation of a central water supply
system, and establishment of a central sewage disposal and collection system.

Polta[86] presented a 1959 discussion of septic tanks as a potential source of nitrogen and phosphorus contamination due to effluent discharge by means of tile fields and seepage pits. Many soils
reduced possible phosphorus contamination by their phosphorus-
fixing capabilities. The extent of nitrogen flow in groundwater was

affected by the adsorption-ion exchange phenomena exhibited by soils, along with the action of nitrifying and denitrifying bacteria. Phosphorus was not seen as a serious contamination threat, but nitrogen in groundwater caused risks of eutrophication and methemoglobinemia.

Hall[87] described a 1968-1970 laboratory study of phosphorus retention by three Maine soil types. Soil column studies were conducted, using both an aqueous solution of known phosphorus concentration and a natural septic tank effluent so that the effects of the soil biota on the retention of phosphorus could be determined. All three soils exhibited a significant capacity for phosphorus retention, but this capacity was not inexhaustible. Therefore, extreme care should be exercised in locating septic tank-drainfield wastewater disposal systems near groundwater resources.

A detailed literature review of septic tanks and their public health and environmental quality influences by Patterson, et al.[88] appeared in 1971. The consistently poor performance of septic tanks indicated that other waste disposal methods were necessary in densely populated areas, and that more rigorous regulation of design criteria, installation, and operation were required in sparsely inhabited areas. The bibliography of the report contained 127 items.

In 1971 Crosby, et al.[89] recorded the findings of a six-year hydrogeologic investigation of pollution hazards involved with the use of septic tanks and drainfields in the Spokane Valley of eastern Washington. Extensive sampling and analysis revealed no evidence of any groundwater contamination.

In 1972 Waltz[90] reported on problems encountered in developing mountain homesites in the Rocky Mountains of Central Colorado. The homesites often required individual water wells and sewage disposal systems, but the septic tank-leach field system generally was not suited for use in the mountainous terrain where soils were thin or missing. Contamination of groundwater from these malfunctioning septic tank-leach fields had become a problem as sewage effluent directly entered bedrock fractures and travelled large distances without being purified. Consideration of geologic conditions in the site selection of septic tanks, leach fields, and wells was seen as a method of significantly decreasing water well contamination in the area.

Baker and Rawson[91] conducted a field study of groundwater quality in the Toledo Bend Reservoir area of Texas in 1972. Twenty test wells were installed down the land-surface slope from septic tank systems, between the septic tanks and the reservoir. In the spring of 1972, coliform density of shallow groundwater samples ranged from 0.0 to 1,800 colonies/ml. At least one sample from 18 of the wells revealed some coliform presence, and at least one sample from 12 of the test wells contained more than 100 coliform colonies/ml.

# CHAPTER V

# INDUSTRIAL POLLUTION

## WASTE DISPOSAL

The problem of underground disposal of industrial wastes and its relation to groundwater pollution was the subject of general studies by a task group of the American Water Works Association[92] in 1953 and by Ives and Eddy[93] in 1968. The task group study assessed the effects of industrial waste disposal on groundwaters, noted increasing threats to groundwaters across the nation, and recommended statutory control measures. Ives and Eddy surveyed underground waste disposal policies and practices in the states, and specifically the pollution problems presented by individual subsurface waste disposal wells. The nature and extent of the pollution problems, as well as treatment methods, were reviewed. In addition, recommended practices and procedures in establishing administrative guidelines for the use of disposal wells were given.

Ulrich[94] in 1955 described a field situation involving high chloride contamination of wells at Massillon, Ohio. Chloride contents rose from 8 ppm to 670 ppm and 1700 ppm in two wells. The cause was under investigation and was believed to be industrial wastes infiltrating from an adjacent river; however, upward movement of deeper saline water was also a possibility.

In a 1954-56 study at Indian Hill, Ohio, Parks[95] reported on brine being discharged from a water softening plant which percolated 850 feet to pumping wells. The chloride concentration in the well waters rose from 20 ppm to 744 ppm. The brine discharge point was moved 1,000 feet farther away, the most polluted of the four wells was pumped to waste, and a large pit was dug in gravel into which river water was pumped. As a result, the chloride content in the wells fell to 26 ppm by 1956.

Waste disposal practices and the resultant contamination of groundwater in the Rocky Mountain Arsenal area of Colorado were reviewed in two significant studies since 1950. In a 1954-56 study, Petri[96] reported on chloride increases in groundwater of the area of up to 200 percent due to infiltration from disposal ponds. The control method employed was to line these ponds to stop the infiltration of wastes. Walker[97] in 1961 described the improper waste disposal practices of spreading basins as being responsible for groundwater contamination through unintentional artificial recharge. The contaminated groundwater was toxic to crops and unpotable for humans. Despite corrective measures attempted, the area of toxicity was expanding as the polluted groundwater migrated laterally.

In 1962 Swenson[98] reviewed the "Montebello incident" and its aftermath. A chemical plant which produced weedkiller in Montebello, California, discharged dichlorophenol into the groundwater. Within seventeen days all city wells were grossly contaminated, and although the plant stopped the waste discharge within thirty days, foul tastes and odors in the well waters persisted for five years up to 1950. The only corrective measure taken was the treatment of the groundwater with chlorine dioxide.

Evans[99] presented a 1965 analysis of the possibilities of breakdown and dangers of industrial waste treatment and disposal facilities. Particular emphasis was placed on the peculiarly diverse nature of industrial wastes and on the effects of deep well disposal and lagoon systems on groundwater quality.

A 1965 study of the water quality in the Fresno-Clovis area was conducted by the California Department of Water Resources.[100] General groundwater quality was excellent, but the groundwater near the Fresno sewage treatment plant, where effluents were discharged on or under the land surface, was of lower quality. It was found that this groundwater might move toward the city if the water table continued to be lowered, and recommendations were made for conservation measures.

In 1967 Price[101] reviewed the contamination of an alluvial aquifer in Keizer, Oregon. Late in 1946, industrial waste from an experimental aluminum reduction plant was dumped into a borrow pit, and at one time the concentration of sulfate in groundwater exceeded 1,000 ppm. Well water samples were analyzed for hardness as the principal indicator of contamination. Although becoming naturally diluted in the immediate vicinity of the borrow pit, during the period 1947-64 the contaminant spread within the aquifer downgradient for slightly more than one mile.

In 1969 Bergstrom[102] discussed specific waste disposal operations in Illinois, including: landfills and dumps, radioactive waste burial grounds, sewage treatment and waste storage ponds, disposal wells, and sewage-stormwater tunnels. Waste management proposals intended to aid in the protection of groundwater centered on waste disposal site selection criteria, hydrogeologic data necessities, and investigations relating to saturation and water movement in typical geologic terrains.

## INDUSTRIAL AND PETROLEUM PRODUCTS

Maehler and Greenberg[103] presented a 1962 report on a special study undertaken to evaluate organic pollutants in groundwaters. The volume and means of disposal of various petroleum industry wastes were analyzed, and the results indicated the value of organic analyses in pollution studies. The wells sampled were grossly polluted with organic compounds, and the organic compounds discovered were clearly related to the compounds in oil field wastes.

In 1962, Deutsch[104] reported on phenols in the groundwater of Alma, Michigan, from 1945 to 1960. Well waters gave off a foul taste and odor, and led to the discovery of refinery wastes being discharged into a pit and percolating to the aquifer. The glacial drift deposits allowed substantial vertical and horizontal migration of pollutants, and two wells were abandoned as a result of deep percolation to the underlying aquifers. To control the contamination, the pit was sealed and pumping was restricted to decrease the hydraulic gradients. Phenol continued to be present in small quantities up to 1959, however, and the possible use of scavenger wells was discussed.

A detailed statistical survey of worldwide pollution due to petroleum products was presented by Zimmerman[105] in 1964. The sources, extent, and effects of petroleum products contaminated were detailed, as were control and detection techniques.

In 1970 Grubb[106] described a 1967 break in an industrial waste discharge line which, coupled with a 49-foot rise in the Ohio River, allowed hydrochloric acid to enter a Pleistocene outwash aquifer used by a Kentucky industry. Chloride concentrations in excess of 30,000 mg/1 were observed in water discharged from the industrial well nearest the break, and within a year that well was abandoned. Fluctuations of chlorides in an industrial well near the river for a sixteen-month period indicated a persistent body of highly mineralized groundwater near the acid source. The movement of this body of water was restricted at low river stages by the inclined surface of the shale bedrock.

Van der Waarden, et al.[107] in 1971 reported on laboratory studies conducted on the transfer of hydrocarbons from a residual oil zone to trickling water. The study attempted to model oil spills in soils, and a pack of nonadsorbing glass particles was used as a soil model to study the general transfer processes of oil components to groundwater. When the glass particles were replaced by natural dune sand, the transfer of oil components was delayed by adsorption and their concentration in the drain water decreased correspondingly. It was thought that adsorption effects under field conditions might be stronger, and that oxidation and evaporation might also determine the fate of oil in soil to some extent.

In 1971 and 1972, both Williams and Wilder[108] and McKee, et al.[109] reviewed the effects of a gasoline pipeline leak near Los Angeles, California. Since 1968, 250,000 gallons of gasoline had seeped into a valuable groundwater supply. Remedial responses included extensive analytical studies of the two-fluid flow system, as well as clean up and restoration attempts. As of 1971, a system of skimming wells had removed about 50,000 gallons of free gasoline from the aquifer.

A survey of petroleum contamination of groundwater in Maryland was presented by Matis[110] in 1971. Most counties recorded some cases of contamination, with the coastal plain region exhibiting the fewest. Throughout the state, however, problems were very localized. Because it was virtually impossible to remove the petroleum products from the groundwater, legal and regulatory problems continued long after the original complaints.

Collins[111] undertook a 1971 review of the pollution potential of oil and gas well drilling. The mechanisms by which brines, crude oil, or gas may infiltrate and pollute groundwaters were discussed, as were current disposal techniques. The conclusions of the review centered on the need for improved control methods and safe, permanent disposal techniques for the residues of oil and gas well drilling.

In 1972, the Committee on Environmental Affairs of the American Petroleum Institute[112] published a survey on the migration of petroleum products in soil and groundwater. General aspects of petroleum products contamination were discussed, and several incidents of oil pollution were listed. The report contained detailed material on controlling oil spills, the recovery of oil, and the detection of hydrocarbons, along with numerous valuable illustrations.

## METAL WASTES

The discharge of plating wastes by aircraft manufacturing plants in Nassau County, Long Island, New York, has been a continuing topic of study since 1950. Davids and Lieber[113] in 1951 reviewed the situation since 1942 and examined the problem of chromium contamination by diffusion wells and shallow pits. All large industrial consumers of chromic acid were required to install treatment facilities to remove hexavalent chromium from their waste streams prior to disposal. The results seemed encouraging since existing treatment facilities were capable of almost completely removing the toxic elements before the wastes were returned to the ground. Groundwater quality was expected to improve because of this treatment, by the constant dilution by recharge from rainfall, and by diffusion as the water traveled through the ground.

However, in 1954 Lieber and Welsh[114] reported on the discovery of cadmium, a heavier and presumably more toxic metal than chromium, in the groundwater of the area. Test wells were constructed and sampled, and the cadmium concentration ranged from 0.01 to 3.2 ppm. The revealed path of the contaminant also coincided with the direction of groundwater flow. The appearance of cadmium as a groundwater contaminant, believed to be unique, created a need to ascertain safe and reasonable limits for cadmium in potable waters, based on the physiological effects of continuous ingestion of minute particles of the substance.

Again in 1962, Lieber, et al.[115] conducted a field study of both cadmium and hexavalent chromium in the groundwater near South Farmingdale, Long Island, New York. Contamination from a metal plating plant since 1942 was summarized, and the present condition was analyzed through the use of ninety observation wells and water samples at five-foot depth intervals. The best methods of control seemed to be improved treatment facilities at the source of the contamination.

A similar study in the South Farmingdale area in 1962 by Perlmutter, et al.[116] monitored the characteristics and movement of a slug of contaminated groundwater beneath a metal plating plant. Test wells were drilled to 140 feet and sampled at 5-foot depth intervals. Concentrations of chromium, determined by use of S-diphenylcarbazide colorimetric comparison, and of cadmium, determined by dithizone extraction, were high, but less than in the past. The slug was diluted greatly as it discharged into a nearby creek, and there was no danger to the nearest public supply wells. Improved treatment procedures were expected to reduce cadmium and chromium concentrations even further.

In 1970 Perlmutter and Lieber[117] reviewed the contamination by plating wastes in the area since 1941 and described the extent of the pollution in the Upper Glacial aquifer. The seepage of plating wastes containing cadmium and hexavalent chromium into the aquifer had formed a plume of contaminated water 4300 feet long, 100 feet wide, and 70 feet thick. After attaining a maximum concentration of 49 mg/1. in 1949, chromium concentrations had decreased to less than 5 mg/1. in most of the plume. No public supply wells had been contaminated by the metal plating wastes. The report also discussed in some detail the factors contributing to the longitudinal, lateral, and vertical spread of the contaminants.

## MINES

A report on the problem of acid mine drainage in the Appalachian Mountain region was presented by Rice and Co.[118] in 1969. Cost curves for various techniques of controlling such pollution were developed. In addition, the effectiveness of control techniques was studied in terms of the degree to which mine drainage pollution was controlled, and the quality characteristics of the post-technique water. The recommended techniques were neutralization, reverse osmosis, streamflow regulation, deep well disposal, land reclamation, revegetation, pumping and drainage, water diversion, mine sealing, refuse treatment, and impoundment of acid water.

Emrich and Merritt[119] reported in 1969 on the degrading effects of mine drainage on groundwater in Appalachia. Oxidation and leaching connected with coal mining produce high iron and sulfate concentrations and low pH in groundwater. Even with cessation of mining, decades are required before the groundwater again becomes usable. In the Toms Run drainage basin in northwestern Pennsylvania, the effects of coal mining and oil and gas well drilling were studied. The oil and gas wells, along with the natural joints and fractures of the rocks, permit acid mine drainage to move downward from strip mines into underlying aquifers, thereby increasing the iron and sulfate content of the water.

Application of the principles of groundwater hydrology to pollution problems facing the mining industry was examined by Moulder[120] in 1970. Costly treatment of acid mine drainage might be avoided by diverting the source of groundwater to an abandoned mine. In addition, the development of a large groundwater supply near an ore deposit might enhance the possibilities of a leaching operation.

In 1970 Ahmad[121] presented a pilot plan intended to solve the serious problem of acid mine drainage pollution of Lake Hope,

Ohio.  The coal mines of the area were continually being flushed by the disturbed natural groundwater flow system and produced sulfuric acid.  Clay layers under the coal in one mine did not allow water to leak into the underlying aquifer.  However, three separate aquifers existed near the Todd mine, and the plan proposed was to discharge the uncontaminated water from the upper to the lower aquifer, thus stopping the flow of water through the mine.

Ahmad[122] also edited the 1971 Proceedings of the Acid Mine Drainage Workshop, dealing with acid mine drainage pollution problems in the Ohio and Appalachian area.  The need for analysis of drainage water samples was stressed, and suggested control methods included neutralization and mine sealing.

Mink, et al.[123] conducted a 1969-70 study of the water quality of the Coeur d' Alene River Basin of Idaho.  Groundwater samples were generally within acceptable limits, and there was no apparent problem from abandoned mine drainage.  Nevertheless, the groundwater in certain areas was high in zinc and, to a lesser extent, lead content.

More particularly, however, a 1969-71 study by Mink, et al.[124] of the Coeur D' Alene District near Wallace, Idaho, revealed high zinc, lead, and cadmium concentrations occurring in the groundwater.  The pollution resulted from the leaching of old mine tailings that were intermixed with the upper part of the sand and gravel aquifer.  The problem was compounded by a nearby settling pond which recharged the groundwater, raised the water table, and caused more leaching.

In 1972 Galbraith[125] presented the results of a field study at Cataldo Mission Flats, near Coeur d' Alene, Idaho.  The leaching of heavy metals by groundwater passing through mine tailings was caused by the oxidation of sulfides through the action of microorganisms.  The report analyzed the chemical processes involved and the relationship between the concentrations of elements and the pH of the groundwater and tailings.

Merkel[126] conducted a resistivity survey in an area of good geologic control to determine if resistivity was a viable technique for delineating aquifers contaminated with acid mine drainage. Based on the results, he presented in 1972 techniques for the periodic monitoring of groundwater through surface resistivity techniques to determine both the extent and degree of acid mine drainage contamination.

## OIL FIELD BRINES

In 1967 a subcommittee of the Interstate Oil Compact Commission[127] published a survey of water problems associated with oil production in the United States. The nature and extent of the problems, as well as current disposal methods, were outlined, and detailed reports on individual states' regulations and enforcement policies were presented.

Also in 1967, articles appeared in *Petroleum Equipment and Services*[128] and *Petroleum Engineer*[129] which dealt with federal, state, and local regulatory agency attempts to control pollution resulting from mishandling of oil field brines. Current disposal techniques discussed included disposal wells, open pits, and lined reservoirs. The most serious mishandling of salt water from oil field brines was due to disposal in unlined pits, a practice which was being outlawed in state after state. New disposal techniques in the field of waste water disposal were also reviewed.

Groundwater pollution from natural gas and oil production in New York was analyzed by Crain[130] in 1969. Leakage of natural gas wells resulted in groundwater pollution in the form of salt water or gas, but the effects were very localized and were difficult to separate from natural contamination. Oil and salt water in the groundwater was due to oil production, especially the secondary recovery of oil by the water flooding method. Pollution from active oil fields was caused by separator units that disposed wastes on the ground, and by well leakage and spillage. Pollution in abandoned fields was usually caused by the upward movement of the contaminants under artesian pressure through uncapped or leaking wells. The groundwater pollution problems from oil production were expected to increase substantially in the future.

In 1970 Bain[131] detailed the brine storage and disposal problem in the Pocatalico River Basin of West Virginia. Salt brine, oil, or gas were present throughout the shallow salt sands of the region. An increase in brine disposal well drilling threatened upward movement of these pollutants to fresh groundwater unless all wells tapping the salt sands were permanently and properly cased.

Groundwater contamination due to oil field brines in Morrow and Delaware Counties, Ohio, was reported by Boster[132] in 1967 and Lehr[133] in 1969. Saline oil field wastes were introduced to evaporation pits, unlined bulldozed pits, and two creeks in the area. Both studies focused on the sources, severity, areal extent, and future movement of the polluted groundwater. Surface resistivity and conductivity techniques supplemented normal chloride analysis, and established the importance of the ion exchange formula which could greatly extend the time period required for natural clearing of saline enclaves. Pettyjohn[134] surveyed the same problems of oil field brine disposal in evaporation pits in Morrow, Delaware, and Medina Counties in 1971, employing shallow observation wells.

The lack of groundwater pollution from oil wells, gas wells, and dry holes drilled in Michigan was analyzed by Eddy[135] in 1965. One exception was in the Saginaw Valley where abandoned wells drilled for salt, coal, and oil remained unplugged. Special emphasis was placed on proper plugging of wells as the most important phase of groundwater protection. The success of Michigan's flexible oil and gas conservation law was also discussed.

Krieger and Hendrickson[136] reviewed the effects of Greensburg oilfield brines on groundwater in the Upper Green River Basin of Western Kentucky for 1950-59. Periodic sampling of about twenty wells and springs revealed that after the 1958 development of the oilfield and the discharge of brine into the groundwater, chloride contents jumped from less than 25 ppm to over 10,000 ppm. The chloride increases were observed as far as 100 miles downstream from the area of heaviest oil production.

A 1960-61 field study by Hopkins[137] in the Upper Big Pitman Basin, near Greensburg, Kentucky, supplemented the Krieger and Hendrickson[136] survey. Deep well injection of the oil field brines was carried out, but abandoned oil and gas test wells allowed the brines to move upward and contaminate fresh groundwater. Potable groundwater was changed to a sodium chloride type with chloride content as high as 51,000 ppm (compared to less than 60 ppm before oil production). The contamination declined as the oil production decreased, but secondary recovery methods threatened even greater pollution if the abandoned wells were not plugged. The author recommended that residual brines be injected into a permeable zone separated from a fresh groundwater zone by impermeable material, and that several wells and springs be sampled periodically for chemical analyses and water level measurements.

In 1963 Wait and McCollum[138] reported on the contamination of fresh water aquifers in Glynn County, Georgia, through an unplugged oil test well. The well was explored with a current meter to 1780 feet. The well penetrated salt and fresh water aquifers and allowed upward migration of salt water resulting in a chloride content of up to 7780 ppm. The chloride appeared to extend 1.5 miles along the hydraulic gradient, and any additional pumping in the area would hasten the contamination of nearby wells. The study stressed the need for better well construction, plugging of abandoned wells, and the location of well fields away from the test well.

Fryberger[139] conducted a 1967-72 field investigation of a brine-polluted aquifer in Miller County, Southwest Arkansas, from an oil field disposal pit and disposal well. A 4.5 square mile area was affected, and the pollution was expected to last 250 years. Attempts at rehabilitation included pumping the brine into the Red River and deep well disposal techniques, but none of the control methods was considered economically justified.

In 1963 Powell, et al.[140] summarized a field study of oil field brines in six Alabama oil fields. Groundwater contamination problems were observed in four fields, with the major sources being disposal pits above permeable sands and leaks from pipelines and well heads allowing brine percolation. A detailed chart was presented for the four problem fields showing (a) identified and/or possible pollution sources and locations, (b) control recommendations, (c) monitoring requirements, and (d) future problems. The control recommendations centered on the sealing, enlarging, or closing of evaporation pits, proper disposal wells, and continual well maintenance to prevent leaks.

Knowles[141] discussed the hydrologic aspects of Alabama oil field brine disposal in 1965. Groundwater contamination occurred in all Alabama oil fields, with the Pollard field serving as a typical example. Disposal of brines and other wastes in the Pollard field occurred by injection into disposal pits or discharge into evaporation pits. Proper lining of the pits and maintenance to prevent brine and oil leaks were seen as effective methods of contamination prevention.

Irwin and Morton[142] in 1969 produced hydrogeologic information on the Glorieta Sandstone and the Ogallala formation in the Oklahoma Panhandle and adjoining areas as related to underground waste disposal. Permits for 147 oil field brine disposal wells had been issued in the area, and increased vertical permeability between the two formations could result in the upward movement of brine under hydrostatic head from the Glorieta sandstone into the overlying fresh water aquifers, particularly the Ogallala.

In 1965 McMillion[143] surveyed the hydrologic aspects of disposal of oil field brines in Texas. The magnitude of crude oil and accompanying salt water production was noted, as were current disposal methods. Some brines were disposed of into unlined earthen pits and seeped or overflowed to pollute fresh water, while more were injected into the subsurface where inadequate well completion methods could constitute a longer range problem than surface disposal. The need for brine pollution control programs with the objective of maximum oil and gas conservation and development was emphasized.

Payne[144] in 1966 centered on the effects of brine injection and other disposal techniques in Texas. Drilling, production, and abandonment problems in oil and gas operations were discussed. In addition, guidelines were offered for design, operation, and control of effective salt water injection systems.

In 1966-67 Burke[145] and Page[146] reviewed Texas Railroad Commission regulations of 1965, imposing strict controls over salt water disposal. The Commission's regulations hoped to: (1) place greater emphasis on proper well completions; (2) encourage controlled subsurface injection; (3) eliminate all earthen pits; (4) strictly enforce plugging procedures; and (5) closely inspect completion techniques of water injection wells. Page[146] also discussed a number of recommendations intended to serve as guidelines for effective and economical subsurface fluid injection and disposal.

Burnitt, et al.[147] conducted a 1957-61 field study of salt water disposal in three oil fields in Limestone County, Texas. Brines were disposed of in large surface pits, and the resulting degradation of groundwater quality was compounded by 600 abandoned oil and gas wells which were improperly plugged. A detailed coring and sampling program was recommended to determine the overall degree of groundwater contamination in the area.

A 1962 study by Burnitt[148] of soil damage and groundwater quality in Fisher County, Texas, revealed serious soil damage problems on 25 farms due to rising water levels and a general increase in groundwater salinity. In some areas, the presence of sodium chloride-type shallow groundwater suggested brine contamination from a great number of bore holes with improper plugging. The report recommended stricter well construction and plugging standards, the discontinuation of unlined surface pit disposal and injection disposal in the area, and a comprehensive water sampling program.

In 1965 Fink[149] reported on chemical analyses of water from water and oil wells near Harrold, Wilbarger County, Texas. Highly mineralized contaminants could come from subsurface disposal of oil field brines, poor casing or cementing in oil wells, inadequately plugged abandoned wells, and seepage from unlined pits formerly used for brine disposal. Possible control methods included prohibition of brine injection into oil wells which permit upward migration, discontinuance of unlined disposal pits, and periodic tracer surveys on all disposal wells.

In 1969 Preston[150] surveyed groundwater occurrence and quality in Shackelford County, Texas. Small amounts of groundwater, used mostly for household needs and livestock watering, were produced from formations of Permian age and from Quaternary alluvial deposits. Oil field brine disposal methods were thought to be the probable cause of some of the poorer quality groundwater of the area.

Rice[151] in 1968 discussed salt water disposal techniques in the Permian basin of Texas and New Mexico. In order to prevent further groundwater contamination, systems of reinjecting produced water into brine bearing formations were analyzed. Various design, installation, and operation factors were considered for safe and economical fieldwide systems.

In 1968-69 McMillion[152] conducted a field investigation of mineralized groundwater from oil field brine pits in Eastern New Mexico. It was suggested that the poor quality groundwater be pumped and used for the secondary recovery of oil. In addition, three corrective measures for the thousands of discontinued brine pits in the region were offered: (1) restrict fresh water pumping in zones where groundwater movement would influence the pollutants; (2) remove the highly mineralized water from the area; and (3) develop and pump the poor quality groundwater so that its movement is held in check.

In 1971 Rold[153] surveyed pollution problems in the "oil patch" of the arid West. The sources of groundwater contamination included evaporation pits, insufficient surface casing, any injection system, abandoned wells, and seismic shotholes. Constant planning, monitoring, and policing of oil field operations were necessary to prevent increases in total dissolved solids and released crude oil in the groundwater.

The problem of contaminated groundwater due to oil field brines in Elm Creek Valley, Barber County, Kansas, was investigated by Williams and Bayne[154] in 1946. The groundwater quality varied according to the location in the valley, and the possibility existed of further salt water encroachment under heavy pumping conditions. The report concluded that pumping tests should be conducted before groundwater supplies are developed in the parts of the valley where the alluvium is thickest and most permeable.

The general fresh water pollution hazards related to the petroleum industry in Kansas were outlined by Jones[155] in 1950 and Latta[156] in 1963. The causes of oil field brines in groundwater included inadequate well casing, abandoned holes with improper or no plugging, and seepage from brine ponds. In 1963 most disposal was by means of disposal or repressuring wells with only a few surface ponds. There was also the added possibility of groundwater contamination due to escaping gas from petroleum storage cavities. The state pollution control regulations appeared adequate to meet the oil production and brine disposal problems.

A 1953-55 field study of the Raisin City oil field in Fresno County, California, by the State Division of Water Resources[157] revealed contaminated groundwater due to oil field waste water disposal techniques. Chloride contents as high as 2680 ppm were detected in area wells. Injection wells seemed to be a satisfactory disposal method, but it was recommended that the discharge of brines to unlined evaporation-percolation sumps be halted.

A 1957 article in the *Oil and Gas Journal*[158] dealt with brine disposal problems in the San Joaquin Valley in California. The worst potential pollution spots appeared to be under control through injection and percolation, but some oil operators were changing or correcting disposal techniques. The objectives of a waste water disposal system, as well as operating costs, were also discussed.

## PITS AND LAGOONS

In 1963 Harmeson and Vogel[159] surveyed the physical, chemical, bacterial, and radioactive pollutants produced by artificial recharge through recharge pits. They also reported on the 1951-59 experience of Peoria, Illinois, in treating river water with chlorine and discharging it into recharge pits. Existing water quality standards were met, but the presence of ABS and radioactivity in the groundwater was seen as posing an increasing threat. The report contained charts of a full range of water quality analyses.

Wichman and Ahlers[160] reported on the methods used by the Brookhaven Laboratory in Upton, New York, to conserve nuclear reactor cooling water by recharge without disturbing the natural groundwater balance. A simple well water recycling system was employed. The used warm water was piped into one of two recharge basins, consisting of ordinary precast drainage domes placed over perforated concrete slabs. With the continuous disposal of the

warm water, no measurable increase of the groundwater temperature had been found from 1962 to 1967. Little maintenance was
required due to daily changeover of the basins.

In 1968 Preul[161] described field observations made over a
three-year period on nutrient concentration in groundwaters near
ten waste stabilization lagoons in Minnesota which treated domestic
wastewater from small municipalities. Subsurface soils were sandy
or silty, and the ponds had high percolation rates. Ammonia nitrogen concentrations were largely adsorbed within about 200 feet of
a pond. Phosphates were not found in significant concentrations,
being adsorbed by soils over a wide pH range. Alkyl benzene sulfonate, however, was found in the range of 0.5 mg/1 as far as 200
feet from a pond, which could be true of other organics resistent
to biodegradation.

A 1970 discussion of lagoon technology and treatment methods by Middletown and Bunch[162] concentrated on the water pollution drawbacks of lagoons and the place of lagoons in the future.
The main threat to groundwater quality from lagoons was seen as
the difficulties involved in sealing them.

According to a 1970 report by Tossey[163] disposal of digested
sludges in sludge lagoons had been an accepted practice in the city
of Dayton, Ohio, for 35 years. The success of sludge lagooning
was found to depend upon the quality of sludge entering the lagoon.
No documented evidence of groundwater deterioration had been
found, but a system of testing was being organized to determine
the current groundwater status.

Hackbarth[164] in 1971 presented a new method for supplement-
ing data from piezometers to monitor waste disposal sites. The
method involved examination of a time sequence of resistivity measurements at fixed points in a disposal area. Spent sulfite liquor
movement away from a seepage pit was studied using this method.
The study also listed the conditions to be met for the method to
provide results which correlate with specific conductivity of water
samples from piezometers.

Wells[165] reported in 1971 on an attempted study of the effect of unlined treated storage ponds on groundwater quality in
the Ogallala aquifer near Lubbock, Texas. The research was focused on nitrate as a pollutant, but did not accomplish its stated objective because the complexity of the problem was underestimated.
Recommendations for a similar successful project included: careful metering of all water utilized; a dense network of observation
wells, not production wells, for sampling and monitoring; analysis

of representative groundwater samples; and additional research on the nitrogen concentration-percolating water quality relationships involved.

A study by Leggat, et al.[166] on the disposal of liquid wastes into unlined pits at the Linfield disposal site in South Dallas, Texas, showed waste materials percolating into the underlying groundwater reservoir. Upon reaching the high water table, the effluent mixed with the groundwater and moved down the hydraulic gradient, eventually to be discharged to the Trinity River. The heterogeneous groundwater quality observed in the test wells resulted from different types and volumes of liquids placed in the pits and from the random cycle used in filling the pits. Waters from all test wells were highly mineralized, and continued use of the pits was expected to result only in further degradation of groundwater quality.

## RADIOACTIVE MATERIALS

In 1957 deLaguna and Blomeke[167] discussed the disposal of radioactive wastes from the processing of solid fuel elements and from solid blanket elements. The report covered several methods of uranium extraction, removal of element jackets, treatment of uranium-zirconium fuel elements, deep well disposal problems, well hydraulics, thermal considerations of disposal aquifers, regional hydrology, potential deep well disposal areas in the United States, and disposal costs.

Roedder[168] in 1959 detailed problems in the disposal of acid aluminum nitrate high-level radioactive waste solutions by injection into deep, brine-saturated aquifers (salaquifers). The concept of a "zone of equilibration" was developed to aid in discussing the mechanics of the interaction of moving wastes with salaquifer minerals. The width of the "zone" controlled the usable storage capacity per well of the salaquifer. It was shown experimentally that reactions occur with carbonates, limonite, clays, and other typical salaquifer materials, most likely causing precipitation of aluminum and ferric hydroxide gels, effectively blocking further injection. It thus appeared that only under certain very special conditions would the injection procedure be economically feasible.

The injection of low- and intermediate-level radioactive wastes into deep geologic formations was seen as a feasible and economic approach by Kaufman, et al.[169] in 1961. Deep injection was considered in general a less satisfactory alternative than disposal by dilution into rivers and oceans, and the necessity of pretreatment (holdup and blending, chemical precipitation, filtration, pH adjustment, and chlorination) of wastes to insure compatibility with the receiving formation was stressed.

Belter[170] in 1963 reported on radioactive waste management activities of the Atomic Energy Commission. Two basic approaches to effluent control were defined: "dilute and disperse," and "concentrate and contain." The role of specific environments in waste disposal practices was discussed, as was the distinction between basic radiation protection standards and performance criteria of control operations. Examples of radioactive waste disposal practices for various types of wastes were also described. Finally, the current status of research in the area was noted, as well as general economic factors relating to radioactive waste handling and disposal.

In 1965 Mawson[171] reviewed the principles and problems of radioactive waste management. His book covered the sources and nature of radioactive wastes, treatment of gaseous, liquid, and solid wastes, various methods of storage and disposal, and monitoring and control problems. Geologic formations recommended for radioactive waste disposal were salt formations, deep wells in fractures produced between bedded strata by high pressure injection, and deep caverns in original caves or mined cavities where no danger of groundwater contamination existed.

Clebsch and Baltz[172] in 1967 examined the basic technology of the petroleum and chemical industries for deep well disposal and of liquid and gaseous radioactive wastes. An understanding of the physical and geologic characteristics of the disposal reservoir, the effects of chemical reactions between waste and reservoir rock, and the hydraulic effects of long term injection on mass transport rate and direction and on the integrity of bordering geologic units was considered essential for safe and effective disposal. The prospects were considered better for injecting gaseous wastes into unsaturated rocks and for routine disposal of waste gases that could be separated as a low-volume stream.

Theoretical aspects of the transport of radionuclides by groundwater from the site of a nuclear detonation to points of potential water use were studied by Lynch[173] in 1964. Analyses of water transport and contaminant movement were made, and equations presented for predicting transport time and dispersion in uniform systems. The equations indicated that dispersion should have a negligible effect on transport time, but observed cases of dispersion in granular rocks suggested that geologic inhomogeneities play an important role.

Stead[174] in 1964 also reported on the distribution in groundwater of radionuclides from underground nuclear explosions. Radioactive waste disposal operations revealed groundwater transport of radionuclides for considerable distances, and the necessity for hydrogeologic analyses of proposed disposal sites was stressed. The report also discussed precautions available to prevent post-explosion movement into groundwater of long-lived and biologically significant radionuclides.

In 1968 Champlin and Eichholz[175] reported on a specific laboratory study of the movement of radioactive sodium and ruthenium. Radioactive solutions were injected into a model aquifer, and the appearance of the radioactivity in the effluent correlated with increases in suspended particulate matter, potassium and calcium concentration, and overall conductivity. Both sides indicated that significant amounts of radioactivity were transported through the test bed on particulate matter, despite the high solubility of the sodium ion used.

Witkowski and Manneschmidt[176] in 1962 described a decision to discontinue the ground disposal of liquid wastes at Oak Ridge National Laboratory, Tennessee. Remote long range possibilities of serious ground and surface water contamination, along with public relations problems, were cited as factors influencing the decision.

The waste disposal facilities of the Savannah River Plant near Aiken, South Carolina, were described by Reichert[177] in 1962 and by Reichert and Fenimore[178] in 1964. The hydrogeologic, climatic, and demographic characteristics of the area did not encourage the disposal of radioactive wastes to the environment. Thus, ground disposal had been limited to the burial of solid wastes and the discharge of very low-level liquid wastes to seepage basins. No radionuclides had been detected in groundwater due to leaching of the solid wastes, but strontium-90 was observed in sand layers up to 500 feet from the seepage basins. Wherever soils did not contain sandy strata or sand filled clastic dikes, radionuclide migration was slow.

In 1965 Proctor and Marine[179] reported on an investigation which established the technical feasibility and the high degree of safety attainable by storage of high-level radioactive wastes in unlined vaults excavated in crystalline rock 1,500 feet beneath the surface of the Savannah River Plant. The most significant force aiding radionuclide migration from the storage site was derived from the natural groundwater movement, coupled with effects due to dispersion and ion exchange.  Three factors prevented radionuclide migration, however:  the very low permeability of the crystalline rock, the virtually impermeable clay layer separating the rock and overlying sediments containing prolific groundwater zones, and the ion exchange properties of the sediments.  Any one of the three barriers could contain the radionuclides longer than the 600 years required to render the wastes innocuous.

In 1971 Gardner and Downs[180] evaluated the Project Dribble site, Hattiesburg, Mississippi, in terms of radioactive waste disposal and groundwater quality.  The fresh water aquifers of the region were found to have very slow rates of movement, and no excessive radionuclide concentrations were forecast.  However, the need for an expanded water quality monitoring program was stressed.  Release of the land for agricultural pursuits was recommended, but drilling and mining restrictions were still suggested.

Lynn and Arlin[181] in 1962 described a deep well injection system near Grants, New Mexico, for the disposal of uranium mill tailing water by The Anaconda Company.  The reservoir sandstones contained water similar to the injected wastewater, and were isolated from overlying fresh water aquifers by an evaporite barrier zone.  The mill tailing water was decanted, filtered, and introduced into the well by gravity at an average rate of 400 gpm.  The life expectancy of the reservoir was put at ten years.

A Los Alamos, New Mexico, disposal area for liquid radioactive wastes was the subject of a 1966 study by Purtyman, et al.[182].  The fine particles in the alluvial materials had a greater affinity for radionuclides than the more abundant coarse particles.  The radioactivity in the alluvium was dispersed by wastewater and storm runoff and decreased with distance from the point of effluent outfall.  Most of the radionuclides were retained in the upper three feet of the deposits, resulting in very little groundwater quality change.

Since 1961 numerous studies have been presented on the disposal of liquid radioactive wastes to a leaching pond at the National Reactor Testing Station in Idaho[183 to 187].  Peckham[183] in 1961 concluded that the saline waste solutions were moving in the direction of normal groundwater flow at average rates of 15-50 feet per day.  Jones and Shuter[184] in 1962 described seepage from the pond since 1957 and the resulting body of perched water on an extensive sedimentary bed about 150 feet underground.  The observed tritium content of the perched water was thought to be much too low, given the annual discharge rates. Morris et al.[185] in 1964 investigated the chemical and radiometric changes that occurred as groundwater containing radioactive wastes moved through the basalt and unconsolidated sediments of the region. Water level and test hole sample analyses were detailed.

In 1972 Schoen[186] reported on a hydrochemical study of the NRTS area groundwater. Thermodynamic analysis of the data indicated the possibility of calcite and dolomite precipitation during utilization of the groundwater or during waste disposal. Subsurface disposal of liquid waste with a high pH could cause rapid precipitation and well plugging.

Finally, in 1972 Nebeker and Lakey[187] surveyed the liquid waste management system of the NRTS test reactor area. Liquid radioactive waste disposal practices, problems, solutions, and proposed system changes were discussed. In addition, detectable concentrations of various radioactive wastes in the groundwater were presented.

Geological and hydrological aspects of the disposal of liquid radioactive wastes at the Atomic Energy Commission's Chemical Separations Plants at Hanford, Washington, have been the general subject of many studies since 1956[188 to 192]. In 1956 Brown, et al.[188] reported on the rate and direction of flow of groundwaters in the area and the effects of disposal operations. Much emphasis was placed on microgeologic and microhydrologic procedures and concepts of prediction. Raymond and Bierschenk[189] in 1957 described the hydrologic, geologic, and radiologic monitoring data obtained from several hundred wells in the area over twelve years.

In 1958 Bierschenk[190] discussed the location, extent, and hydraulic characteristics of groundwater mounds created by the infiltration of large volumes of radioactive effluents. The natural hydraulic gradients had been reversed in certain locations and migration rates had increased. Such data were valuable in determining the most effective placement of monitoring wells to permit prediction of low-level radioactive waste behavior in the zone of saturation.

In 1959 Bierschenk[191] also reported on general aquifer characteristics and groundwater movement at Hanford. Large amounts of intermediate-level radioactive wastes had been discharged to the ground, and ten times as much (35 billion gallons since 1944) uncontaminated process cooling water had been discharged into open ponds or swamps. The semiarid climate, permeable surficial sediments, and the deep water table formed a situation wherein most of the radioactive materials in the waste were trapped by electro-chemical bonds in the sediments during percolation. Those wastes that reached the water table moved with the groundwater, but their path and concentration depended largely on heterogeneity and anisotropy of the aquifer, and the dispersal of the wastes in the groundwater.

Brown and Raymond[192] in 1962 described methods used at Hanford to measure geohydrologic features affecting radioactive waste disposal. Basic concepts discussed included vertical ground-water movement, dispersion of contaminants, aquifer anisotropism, groundwater flow rates, and the hydrologic continuity between well and aquifer.

Also in 1962 Brown and Raymond[193] summarized Hanford's radiologic monitoring program. Significant increases in groundwa-ter flow information and improved equipment and methods for determining radiocontaminants in the groundwater were described.

In 1964 Raymond and McGhan[194] employed scintillation well probes to monitor subsoil contamination beneath the Hanford ground disposal facilities. Results indicated that significant lateral spread of radioactive wastes occurred in the sediments of the study region, and that the downward migration rates of gross gamma emitters was relatively slow.

Brown[195] in 1967 detailed the migration characteristics of various radionuclides through the Hanford sediments from informa-tion gathered through an extensive network of monitoring wells and sophisticated monitoring equipment. Sediment samples obtain-ed by core drilling showed that over 99.9 percent of the long-lived radionuclides were contained within the first ten meters of the sixty-meter partially saturated sediment column underlying the dis-posal facility. Three radionuclides (ruthenium-106, technetium-99, and tritium) were traced in the groundwater for distances up to fifteen miles, but at only 2½ miles from the disposal sites all radio-nuclide concentrations were below established drinking water limits.

From January 1968, to June 1971, the radiological status of the groundwater beneath the Hanford Project was summarized every six months (with the exception of June-December, 1968)[196] [to 201]. Between 300 and 500 wells were employed for surveillance, and the reports summarized beta, ruthenium, tritium, uranium, and nitrate ion concentrations in unconfined and confined aquifers of the area.

In 1971 LaSala and Doty[202] presented a preliminary evalua-tion of the hydrologic factors related to the feasibility of storing high-level radioactive wastes in deeply buried basaltic rocks at Hanford.  Key factors included the rate and direction of local groundwater movement, the characteristics of groundwater discharge, and the geochemical nature of the waste-rock-water system that might affect radionuclide movement.  A test well was drilled in 1969 to about 5,600 feet, but hydraulic testing was not completed due to the caving of well sections.  However, under prevailing head relationships, and assuming observed geohydrologic conditions were widespread, it was considered feasible to store radioactive wastes safely in mined cavities in thick impermeable rock layers below 1,200 feet.

# CHAPTER VI

# AGRICULTURAL POLLUTION

## AGRICULTURAL WASTES

In 1967 Stewart, et al.[203] analyzed cores from fields and corrals in the Middle South Platte Valley of Colorado to determine the distribution of nitrates and other water pollutants in the area. Significant quantities of nitrate were found in most cores from irrigated fields with row crops or cereal grains, as opposed to low nitrate content in cores from irrigated alfalfa fields. The authors concluded that much of the nitrate under feedlots probably will never reach the water table due to denitrification. Amounts of nitrogen as nitrate found under corrals varied from almost none to over 5,000 pounds/acre in a twenty-foot profile. Large amounts of organic carbon and ammonia were discovered in water samples beneath several corrals, and bacterial counts under corrals were considerably higher than under other areas. The findings indicated some pollution of groundwater by deep percolation was occurring from corrals, but further study was recommended to determine its magnitude.

Again using analyses of soil profiles, Stewart, et al.[204] in 1968 detailed the contributions of fertilizers and livestock feeding wastes to groundwater pollution in the Middle South Platte River Valley in Colorado. Amounts of nitrate in profiles varied widely with land use, and the results were summarized. Feedlots located near homesteads had a much greater effect on nitrate content of domestic well water than did cropped land.

Robbins and Kriz[205] presented a 1969 survey of the relation of agriculture to groundwater pollution. In this survey paper, with 97 references, various agricultural sources of pollution were reviewed, including animal wastes, fertilizers, pesticides, plant residues, and saline waste waters. Included also were different types of solutions to pollution control problems.

A 1969 paper by Smith[206] dealt with the contribution of fertilizers and livestock feeding operations to groundwater pollution. Many shallow wells in Missouri were contaminated with nitrates as a result of leaching from livestock feeding operations. The percentage of nutrients applied in chemical fertilizers moving into the groundwater was thought to be relatively small, and good fertilization practices were thought to lessen nutrient losses to below that lost on unfertilized soils.

Biggar and Corey[207] in 1969 presented a comprehensive review of numerous aspects in the relationship of agricultural drainage to water eutrophication. Particular attention was given to the chemical reactions undergone by nitrogen and phosphorus in the soil-water system. The fate of nutrients transported by deep percolating water was analyzed, and illustrations of plant nutrient loss from harvested areas and contributions of fertilizing elements from agricultural lands were included.

A report by Moore[208] in 1970 discussed the water geochemistry of the Hog Creek Basin in central Texas. The field study described water quality changes in the upper part of the basin along a nineteen-mile reach of the stream. Analyses were made of water samples from each of the major rock formations in the area. The Edwards Limestone, a shallow aquifer, was subject to pollution from agricultural fertilizers, as revealed by unusually high nitrate concentrations.

The movement of agricultural pollutants in groundwater was the subject of a 1970 discussion by LeGrand[209]. In it he concluded that sufficient safeguards were available to minimize groundwater pollution to the extent that good agricultural practices should not be deterred. The unsaturated zone above the water table attenuated almost all of the foreign bodies that were potential pollutants of the underlying groundwater. Environmental factors tending to reduce the pollution of groundwater from wells and springs were presented. There were: (1) a deep water table which allowed adsorption, slowed subsurface movement of pollutants, and facilitated oxidation; (2) sufficient clay in the path of pollutants to favor retention or sorption of pollutants; (3) a gradient beneath a waste site away from nearby wells; and (4) a great distance between wells and wastes.

An essay by Viets[210] in 1971 focused on the many proposals for restricting fertilizer use because of the resultant leaching of nitrogen and phosphorus to the groundwater. He maintained that the data were too scanty and the problem too complex to immediately blame fertilizer use for many pollution problems. He pointed out alternate sources of groundwater pollution in agricultural areas, including sewage, animal wastes, and irrigation. He recommended taking cores from the land surface to the water table and analyzing them for nitrate and rate of water movement before a specific fertilizer restriction could be justified. In general, the author doubted that widespread restrictions on fertilizer use would improve groundwater quality enough to compensate for the risk of a less abundant, more costly food supply.

## ANIMAL WASTES

In 1969-70 Resnik and Rademacher[211, 212] presented an overview of the causes and effects of animal waste runoff. Since feedlots have been located without regard to soil inventory and topographic characteristics, high BOD waste runoff was common. The infiltration of nitrates from manures to well waters was also well documented. The extent of the problem and the present status of regulatory legislation were discussed, along with additional legislative proposals.

In addition, Rademacher and Resnik[213] put forward a model profile for action in 1969. The essential elements of the program were re-education, research, and regulation. Emphasis was placed on proper feedlot location and research devoted to the institutional problems of animal waste management. Animal waste disposal problems required an organized, coordinated, interdisciplinary approach.

In 1970 Miner and Willrich[214] also discussed the pollution potential of animal wastes. Livestock operations and field-spread manure were seen as prime sources of pollutants, and controls through proper animal waste management were examined. In the same publication, McCalla, et al.[215] detailed the possibilities of excessive mineralization of animal wastes in soils resulting in the leaching of nitrate to the groundwater and nitrogen and phosphorus runoff.

Given the nature and extent of the groundwater pollution problem caused by animal wastes, much work has been done on possible methods of control. In 1967 Loehr[216] analyzed the quality of liquid and solid effluents from anaerobic lagoons treating feedlot wastes. Even under ideal equilibrium conditions, the liquid effluent from such lagoons constituted a serious groundwater pollution threat. However, when used in combination with subsequent treatment units, anaerobic lagoons could be an effective process for treating livestock and feedlot wastes that have a high solids content.

Webber and Lane[217] presented a 1969 discussion on the nitrogen problem in the land disposal of liquid manure. They outlined the cropland requirements for the utilization and disposal of nitrogenous compounds. The land spreading objectives were to achieve optimum use-efficiency application rates and to insure that the application rates achieved disposal ends without contributing to environmental pollution. Recommendations were given on how much land was required for crop utilization and pollution control for various livestock operations.

In 1970 Overman, et al.[218] reported on the effectiveness of a soil plant system in renovating waste water from farm animal operations. Plots of ground were seeded with oats, and wastes from 160 cows were applied up to one inch per week. Weekly chemical analyses to a depth of 60 cm. for nitrate and orthophosphate content showed that nitrogen and phosphorus removal was greatly enhanced by plant growth. It was even suggested that a more intense application rate could be used.

Data on the quality and quantity of runoff from beef cattle feedlots were presented by Loehr[219] in 1970. Because of the intermittent nature of this runoff, minimum drainage control was possible using retention ponds. Resulting groundwater pollution problems were briefly discussed.

In 1971 Fogg[220] reviewed the criteria for an effective animal waste management system. A proper system should: (1) divert clean water from livestock waste areas; (2) provide controlled drainage or runoff from such areas; (3) prevent leaching of contaminants; (4) collect polluted runoff; and (5) treat or safely dispose of collected runoff. Solid manure should be removed and stockpiled until it can be safely spread on or deposited in the land. Liquid manure could often be disposed of by a water spreading or irrigation system utilizing the soil and plant cover for treatment, sometimes preceded by the use of aerobic or anaerobic lagoons.

Concannon and Genetelli[221] reported in 1971 on a study of four specific methods of disposing of organic manures which utilized soil as the ultimate disposal media. Lagooning, sanitary landfilling, sub-soil injection, and the PFC method all posed possible groundwater pollution dangers due to heavy loadings of organic and inorganic materials. Chemical and bacteriological analyses were performed for four loading rates of dry poultry solids in field plots and in laboratory soil columns. Total organic carbon concentration levels were high, nitrate and sulfate concentrations exceeded USPHS limits, and all fecal coliform tests were negative. No significant difference was observed between laboratory and field results, lead-

ing to the conclusion that soil columns were an effective controlled means of studying the soil as a disposal media for solid waste.

In 1971 Viets[222] discussed the problems engineers face in de-
signing feedlot facilities that minimize runoff or dispose of it economically and beneficially. Groundwater pollution resulting from
returning the solid waste to the ground was seen as a predominant-
ly local phenomenon. Since only about 10 percent of the land
needed to produce foodstuffs for cattle was needed for productive
waste disposal, zoning was seen as one of the best solutions to the
feedlot problem.

Numerous studies have also been done on specific animal
waste pollution problems in particular localities. In 1967 Stewart,
et al.[223] reported on an investigation of nitrate pollution of groundwater in the South Platte Valley of Colorado, an area intensively
farmed with many concentrated livestock feeding operations. The
average total nitrate-nitrogen content in soil profiles for various
kinds of land use was reported. Groundwater samples often contained high concentrations of nitrate, and those obtained beneath
feedlots contained ammonium nitrogen and organic carbon. The
data revealed that nitrate was moving into the groundwater supply
under both feedlots and most irrigated fields, excluding alfalfa.

In 1969 Evans[224] detailed research on pollution abatement and
management of organic wastes from cattle feedlots in northeastern
Colorado and eastern Nebraska. Feedlots had the highest nitrate
levels, but irrigated land probably contributed more total nitrate to
the  groundwater due to its much larger acreage. A rapid dying of
the coliform population in feedlot soils indicated little danger of
groundwater contamination by coliforms.

In 1970 Mielke, et al.[225] reported on the groundwater quality
in the proximity of a level feedlot on a permeable soil with a fluc-
tuating high water table in the Platte River Valley of Nebraska.
Six observation wells, six water level measuring wells, and two recording wells were employed in the investigation. Soil cores were
taken to determine the quantity of nitrate which could move into
the water table. Core sample analyses indicated that downward
movement of nitrates and other forms of nitrogen in the soil was
minor. The 12-15 inches of manure pack decreased the actual penetration depth of the nitrogen into the profile.

Gilbertson, et al.[226] discussed runoff, solid wastes, and nitrate
movement on beef feedlots in Nebraska in 1971. It was found
that runoff quality and quantity depended more on rainfall than
on slope or cattle density, but high density lots yielded about 150

percent more winter runoff than low density lots.  After one year
nitrate movement in soil was minimal.

In 1972 Lorimor, et al.[227] reported on a field investigation of
nitrate concentration in groundwater beneath a beef cattle feedlot
in Central City, Nebraska.  Daily sampling of wells near the feedlot
revealed that the start of irrigation pumping resulted in no significant increase in nitrate levels.  The levels were found to be well below the USPHS limit.

In addition to these Colorado-Nebraska studies, a 1969 overview of the problem of animal waste pollution by Rademacher[228]
reported that of 6,000 groundwater samples analyzed in Missouri,
42 percent contained more than 5 ppm nitrate as nitrogen.

Gillham and Webber[229,230] examined a case of nitrogen contamination of groundwater by barnyard leachates in 1969-70.  From
piezometric potential and hydraulic conductivity measurements,
quantitative flow nets were drawn permitting groundwater discharge
calculations.  During a five month study period, 4.4 pounds of inorganic nitrogen from the barnyard was contributed to the groundwater.  The concentration of nitrogen was related to the direction
of groundwater flow and was dependent on the presence of conditions suitable for leaching and the dilution potential of the local
groundwater flow system.

In 1970 Frink[231] presented analyses of nutrient cycling on
dairy farms in the Northeast showing that significant quantities of
nitrogen may be lost to groundwater.  Calculations of the efficiency
of nitrogen conversion on these farms revealed that losses to the
environment increased dramatically as farm size decreased.  It appeared that a decrease in the total nitrogen imported onto the smal-
ler farms would not seriously reduce productivity.  In addition,
nitrogen loss could be reduced by foliar applications to the growing
crop, selection of varieties with high yield and nitrogen content,
increased plant populations, and more extensive use of cover crops.

In 1971 Miller[232,233] reported on detailed field and laboratory studies on infiltration rates, nitrate distribution, and groundwater quality beneath cattle feedlots in the Texas High Plains.  Infiltration of feedlot liquid waste to the water table below feed-
yards was insignificant in most localities.  Infiltration of feedlot
runoff and subsequent concentration of dissolved ions in groundwater were dependent on, among other things, surface and subsurface geology, depth to water, thickness of the groundwater zone,
and differences in the lateral and vertical permeabilities of the Ogallala formation.  No direct correlation of groundwater quality exist-

ed with feedpen-runoff slope, cattle load, or surface area ratios of drainage basins to collection systems. No regional subsurface pollution problem from cattle feedlot runoff was found to exist, nor was one foreseen.

Crosby, et al.[234] in 1971 analyzed a test drilling program at a dairy in the Spokane Valley of Washington to study the effects of feedlot operation on groundwater quality. Coliforms were found to disappear within a few feet of the ground surface, but chlorides and nitrates were persistent in depth and could actually reach the groundwater. The low natural moisture content of the soil, coupled with apparent high moisture tensions, suggested that soil moisture was not presently moving downward in the system. It was concluded that formation of organic matters in near surface layers would arrest the downward migration of inorganic chemicals from the feedlot environment in time.

Adriano, et al.[235, 236] conducted a 1969-71 field and laboratory study on the fate of nitrate and salt from land disposal of dairy manures in the Chino-Corona Basin near Los Angeles, California. Soil and water samples were taken from sites representing corrals, irrigated croplands, and pastures used as disposal areas. Considerable amounts of $NO_3^-$ and salt were found in soil profiles beneath the disposal areas, although the magnitude was not as high as in profiles under corrals. Average $NO_3^-$-N concentrations in groundwater samples generally exceeded the USPHS recommended limit of 10 ppm for safe drinking water. It was suggested that a reduction in cow population from ten to three per acre would keep $NO_3^-$-N levels in soil within acceptable limits. In addition, maximization of $NH_3$ volatilization from manure before incorporation into the soil was thought to increase the chances for a reduction in $NO_3^-$-N content in soil.

In 1972 Hutchison, et al.[237] summarized a research project conducted in Maine to determine the maximum acceptable rates of manure application in an excessively drained glacial outwash, a well-drained glacial till, and in poorly drained Maine soils. Using field plots and a lysimeter study, results indicated safe nitrogen application rates of 350, 1,400, and 200 pounds per acre for Windsor loamy sand, Charlton fine sandy loam, and a poorly drained Scantic silt loam, respectively.

## IRRIGATION RETURN FLOWS

Oahu, Hawaii, has been the focus of several studies on the effects of irrigation on soils and groundwater. In 1962, Mink[238] reported on an increase of silica and nitrate in the groundwater beneath heavily irrigated sugar cane. The contamination was due to percolation of nitrate fertilizer and the leaching of silica through water-logged soil.

Tenorio, et al.[239] summarized the results of a 1967-69 investigation of the physical and chemical characteristics of irrigation return water in Pearl Harbor-Waipahu and Kahuku, Oahu, and central and west Maui. Well samples, profile samples, and composite samples were obtained in areas used for tropical agriculture. The well waters were evaluated according to Visher and Mink's index constituents (silica, sulfate, and nitrate) and other significant ionic compositions. Analysis indicated a cyclical trend in concentrations of major constituents, either related to seasonal rainfall or irrigation practices, or both.

Tenorio, et al.[240] reported on phase III of the previous study in 1970. Basal water quality of aquifers in Kahuku, Oahu, and Kahului and Lahaina, Maui, were examined, and the effects of prevailing agricultural practices on groundwater were discussed. The presence of irrigation return water indices in groundwater was traced to both fertilization and heavy pumping and recycling of the basal water.

In 1970 Leonard[241] presented a paper on the effects of irrigation on the chemical quality of ground and surface water in the Cedar Bluff Irrigation District of west-central Kansas. One hundred observation wells were monitored, and the chemical quality of the groundwater was found to vary from well to well. Calcium, sulfate, and bicarbonate ions dominated, and the chloride content was found to increase as the irrigation continued. The data suggested that the original groundwater in the district was being diluted and displaced by irrigation water.

Law, et al.[242] in 1970 analyzed the degradation of water quality in irrigation return flows. The study centered on the increase in total dissolved solids in percolating soil water and on the salinity status of a saltwater-irrigated clay loam soil. In general, draining and percolating waters were found to adversely affect groundwater quality. In particular, percolating irrigation water transported about ten tons of salt per acre-foot.

Alfaro and Wilkins[243] reported on a 1970 laboratory model study on salt distribution and effluent concentration in soil profiles. Results indicated that modeling profiles according to the design conditions specified by the theory may be useful in predicting quality changes of irrigation return flows.

In 1971 Thomas, et al.[244] described the development of a hybrid computer program to predict the water and salt outflow from a river basin in which irrigation was the major water user. A chemical model which predicted the quality of water percolated through a soil profile was combined with a general hydrologic model to form the system simulation model. The model was tested on a portion of the Little Bear Basin in northern Utah, and its successfully measured simulated measured outflows of water and of each of six ions for a two-year period. The only discrepancies were in predicted values of small concentrations of sodium ions, which comprise only 2 percent of the total salt outflow. Preliminary results indicated that the available water supply could be used to irrigate additional land without unduly increasing the salt outflow from the basin. With minor adjustments, it was thought the model could be applied to other areas.

In 1971 Law[245] presented the status of the National Irrigation Return Flow Research and Development Program. Current research projects were discussed, along with a number of potential control measures. Improvements in the water delivery system, on-the-farm water management, and the water removal system were considered with respect to improving the quality of irrigation return flows and decreasing the degradation of receiving waters. The need for research and field investigations to evaluate the effectiveness of potential control measures was stressed.

Fitzsimmons, et al.[246] summarized results of a 1970 field study in the Boise Valley of southwestern Idaho. Inorganic materials were discovered in surface- and groundwater in the intensively farmed, gravity irrigated area. The groundwater contained more nitrate-nitrogen (4.92 ppm) than other water sources, perhaps due to leaching of percolating irrigation water, or from feedlots, dairies, and septic tank drain fields in the area. The groundwater also contained relatively large concentrations of both ortho- and total phosphorus (.11 and .58 ppm, respectively), a surprising discovery since it was generally assumed phosphorus was not readily moved through soil by flowing water.

## PESTICIDES AND HERBICIDES

In 1962 Bonde and Urone[247] summarized the results of a study of over 225 wells in Adams County, Colorado. Plant toxicants (chlorate and a toxicant with effects similar to 2,4 dichloro-phenoxyacetic acid) were found in the groundwater of almost one quarter of the wells sampled. All contaminated wells were north-west of the Rocky Mountain Arsenal waste disposal basins in the direction of groundwater flow. Chemical, x-ray, and bioassay tech-niques were employed to identify the chlorate; high concentrations of sodium chloride were found to coincide with toxicant presence.

Two studies have been done on the general adsorption and mobility characteristics of pesticides in soils. McCarty and King[248] found a positive correlation between the extent of adsorption and the clay content of the soils, and an inverse correlation between the extent of adsorption and the rate of pesticide movement. They emphasized that both adsorption and degradation effects had to be considered in predicting the leachability of pesticides in soils. Hug-genberger, et al.[249] presented a mathematical model to predict the distribution of pesticides in a soil profile through the use of an "ad-sorption coefficient", but concluded no accurate prediction could be made of the depth of maximum pesticide concentration.

Eye[250] conducted research on the problem of aqueous trans-port of dieldrin residues in soil. He concluded in 1968 that the adsorptive capacity of soil to dieldrin was so great that penetration through soil was negligible, and no threat of groundwater pollution existed.

A laboratory study on the adsorption of lindane and dieldrin on natural aquifer sands from Portage County, Wisconsin, by Bou-cher and Lee[251] in 1972 corroborated Eye's[250] theory. After three successive washes of distilled water, less than 20 percent of the dieldrin adsorbed by the aquifer sands was removed. However, nearly 70 percent of the adsorbed lindane was leached after similar washes.

In 1969 Robertson and Kahn[252] reported on four experiments of aldrin (a representative member of the chlorinated hydrocarbon insecticide group) infiltrating through columns of Ottawa sand. They concluded that the penetrability of chlorinated hydrocarbon insecticides through soils was dependent upon the type of formula-tion applied, the frequency of its application, soil conditions, and the frequency and rate of rainfall or irrigation.

Dregne, et al.[253] studied the movement of 2,4 dichlorophen-oxyacetic acid (2,4-D) in three soils to determine the extent to which herbicides applied in the field enter the surface and ground-water systems. Primary emphasis was placed on the effect of ex-changeable cations on 2,4-D movement. A variety of analytical techniques indicated that 2,4-D in the salt or acid form was only slightly adsorbed by soil particles. The ease of 2,4-D leaching was found to depend on the relative permeability of the soils.

In 1971 Mansell and Hammond[254] detailed a further experiment on the influence of physical and chemical soil properties upon the transport of 2,4-D and paraquat in columns of organic and sandy soils. Miscible displacement of aqueous solutions of these herbicides through columns of Everglades mucky peat resulted in most of the 2,4-D and all of the paraquat being adsorbed. Similar thorough removal of the herbicides was observed in the fine sands, although the presence of large concentrations of potassium chloride in the soil solution decreased the quantity of paraquat adsorbed. A mathematical transfer function theory was used in connection with statistical hydrodynamics to develop a technique for analysis and prediction of herbicide elution from soil columns during miscible displacement experiments.

In 1967 Johnston, et al.[255] reported on the type and quality of insecticide material found in irrigated agricultural soils in the San Joaquin Valley of California. Relatively small quantities of chlorinated hydrocarbon residues were found in the tile drainage effluent, but higher concentrations were found in effluent from open drains where both surface and subsurface drainage waters were collected. Effluent samples from seven tile drains and samples of applied water and tailwater contained about ten times the amount of residue as the applied water when DDT was used and 85 times as much when lindane was used. Large concentrations of residue were found in the surface soil although there was no direct application.

A 1970 report of the Working Group on Pesticides[256] examined the problem of ground disposal of pesticides and the extent of resulting well and groundwater contamination. Types of pesticide wastes were discussed, and the interactions between pesticides and soils and groundwater considered. Criteria were provided for establishment of guidelines on pesticide waste disposal practices and monitoring. The magnitude of the threat to groundwater was dependent on the properties of the pesticide waste, the hydrological characteristics of the disposal site, and the volume, state (liquid or solid), and persistence of the waste. Particular emphasis was placed on the application of the technology of groundwater occurrence and movement to these problems. Nine specific recommendations on the problem of pesticide waste disposal were offered.

Schneider, et al.[257] in 1970 described an experiment designed to study the movement and recovery of herbicides in the

Ogallala aquifer at Bushland, Texas. Water from an irrigation well was used to inject three common herbicides (picloram, atrazine, and trifluralin) into a dual purpose well. The well was then pumped long enough to recover essentially all of the recharged water. Nitrate was used to trace the movement of the recharged water. Water samples pumped from observation wells at radial distances of 30 and 66 feet from the dual purpose well showed that the herbicides moved through the aquifer with the recharged water. Coliform bacteria and DDT were effectively filtered or adsorbed by the fine Ogallala sand.

In 1971 Olsen[258] surveyed problems in both natural and artificial groundwater recharge employing surface water that had been subjected to mild pesticide contamination. Two specific examples of the groundwater contamination effects of such recharge were discussed, one in Colorado and one in Texas.

Swoboda, et al.[259] in 1971 summarized research on the distribution of DDT and toxaphene in Houston black clay in three watersheds at Waco, Texas. Soil samples indicated that some of the DDT was not adsorbed by the clay and moved downward with water.

In 1971 Lewallen[260] reported on a 1967-71 field study of pesticide contamination of a shallow bored well in the southeastern coastal plains. Pesticide-contaminated soil had been used as a backfill around the well casing. Water, sediment, and soil samples were taken. The contamination of the well water had remained relatively low, probably because of the very low solubilities of the pesticides (DDT, DDE, and toxaphene) in water. The contamination of the well actually occurred through the movement of surface soil containing adsorbed pesticides to the water table.

Dixon[261] collected four detailed studies on the adsorption and decomposition of pesticides (amiben, diquat, endrin, dieldrin, and aldrin) by clay minerals and soils in the same southeastern United States area. The soils of the region were acidic, contained vermiculite and kaolinite as major clay minerals, and had high percolation rates. The studies suggested that these soils were more effective in decomposing organic molecules than neutral or alkaline soils and were less likely to permit leaching of organic ions into groundwater than less weathered soils.

# CHAPTER VII

# POLLUTION FROM WELLS

## DISPOSAL WELLS

Live Oak and Orlando, Florida, were the sites of 1948 investigations by Telfair[262] to evaluate the effect of diffusion of surface drainage, sewage, and trade wastes through drainage wells into the permeable Eocene limestone aquifers. The report gives the results of bacteriological findings, as well as summaries and conclusions concerning pollution and its effects, possible remedies, and future prospects.

In 1952 Reck and Simmons[263] reported on groundwater in the Buffalo-Niagara Falls region of New York. They found that quality was generally good; however, large sections of the Onondaga Limestone aquifer had been polluted by individual and industrial wells drilled for the discharge of waste materials. Many of these wells became clogged, losing their efficiency to absorb waste. Health officials discouraged the practice of drilling drainage wells.

More recently, a 1968 study by Sceva[264] concentrated on drilled disposal wells in the Middle Deschutes Basin in Central Oregon. The Basin is underlain by basaltic lava flows that restrict the construction of conventional drain fields for liquid waste disposal. Large quantities of groundwater beneath this region were threatened by the liquid waste injection and by the construction of deep uncased water wells. Recommendations included the prevention of further drain well construction and the casing of all deep water wells.

Abegglen, et al.[265] have also studied the effects of drain wells on groundwater quality of the Eastern Snake River Plain aquifer of Southern Idaho, the principal domestic water supply resource in the area. Some 3,000 drain wells in Lincoln, Jerome, and Gooding Counties extend into fractured basalt aquifers, and were being used for the disposal of sewage, street drainage, irrigation excess water, and industrial wastes. A bacterial pollution problem existed on a local scale, and corrective measures were necessary to protect the public health in several areas of the Plain. Effective alternatives to the use of drain wells include municipal sewerage, above-ground and subsurface soil absorption systems, and sedimentation-recirculation systems.

## INJECTION WELLS

Groundwater pollution problems related to the subsurface disposal of liquid wastes by means of deep well injection have been reviewed in detail in the literature since 1950.

Warner[266] in 1965 concluded that deep well injection was technically feasible and, if properly planned and implemented, a safe method for liquid waste disposal. Areas of further research needed were outlined. In 1967 the results of a comprehensive study on injection wells and industrial waste disposal by the Interstate Oil Compact Commission[267,268] were presented. Treatment methods, compatibility of fluids and rock, geological aspects, injection pressures and rates, and legal considerations were reviewed. Current injection well systems were summarized, and guidelines established for well applications, drilling practices, monitoring, and well abandonment.

Walker and Stewart[269] and Talbot[270] in 1968 reviewed state deep well disposal practices and regulations. The necessity for a suitable disposal stratum and a waste physically and chemically compatible with the resident material in the disposal formation was stressed. In 1969 Manning[271] detailed similar requirements and suggested that areas underlain by sedimentary rocks were potential disposal reservoirs due to generally large areal formations. Caswell[272] in 1970 also reviewed the technology, hydrology, and legal status of deep disposal wells, and warned that in many cases injection was not feasible due to long-lived wastes or to the hydrogeology of the disposal horizon. In 1972 Cook[273] collected and edited 37 studies on deep well waste disposal and related subjects.

All types of water desalinization schemes have the problem of concentrated brine disposal, and Manning[271] in 1969 suggested that injection wells might provide safe and convenient disposal. Boegly, et al.[274] reviewed the literature on this problem in 1969, and found deep well injection was technically feasible if satisfactory pretreatment was provided. A suitable site for such injection required a permeable sedimentary formation capped by an impermeable formation. This study, as well as Rinne[275] in 1970, stressed the need for detailed geologic and hydrologic investigations in insure site suitability and proper system design.

In 1967-69, Warner,[276,277] *Water Well Journal,*[278] and *Environmental Science and Technology*[279] summarized data on 110 injection wells in use mainly in North-Central and Gulf Coast areas. Some of the characteristics of industrial waste injection wells reviewed were:  operation, location, well depth, depth of injection horizon, geologic formation, chemical and physical character of waste, injection pressure and rate, and information sources.

The cost factors involved in deep well disposal were analyzed by Selm and Hulse[280] in 1960.  The nature of pretreatment required, the depth of the hole, waste corrosivity, state regulations, geology of the formation, and many other considerations were discussed.  The presence of intolerable amounts of dissolved salts was considered mandatory before deep well disposal was definitely more economic than surface disposal techniques.  Stewart[281] in 1968 concluded that in general the cost of deep well disposal was about one-third of any other method of waste neutralization.  In 1969 Manning[271] suggested that the great expense of injection disposal made it best suited for disposal of relatively small quantities of particularly noxious wastes.  Also in 1969 Boegly, ct al.[274] reported on the costs of deep-well oil-field brine disposal systems.

Problems of design, control, and monitoring of deep well injection systems have also been examined.  In a 1966 laboratory and theoretical study, Warner[282] linked the amount of reaction between injected and interstitial solutions to the dispersive character of the porous medium.  The concept of a buffer zone of nonreactive water between injected waste and aquifer water was also proposed.

Basic design principles for disposal well systems were presented by Marsh,[283] Stewart,[281] Walker and Stewart,[269] Talbot,[270] Slagle and Stogner,[284] Rima,[285] and McLean[286] in 1968 and 1969. Design requirements for aquifer protection included:  selection of a zone bounded by aquicludes; strict drilling, casing, and sealing procedures; waste quality and application rate controls; proper surface injection and treatment equipment; stand-by wells; knowledge of hydraulic gradients and hydrodynamic dispersion factors; and a comprehensive system of monitoring wells.  Talbot[270] and McLean[286] also stressed the need for preconstruction testing.  Injectivity tests of the formation's hydrologic properties were urged, and a method for calculating the radius of injection capacity of the formation was described.

The effects of deep injection are complex, and some geologists have felt that the little information available was misunderstood or misapplied.  Sheldrick[287] in 1969 summarized the criticism of the geologic criteria on which feasibility and safety of injection wells were evaluated. Knowledge of the hydrodynamics of underground formations and underground monitoring techniques were thought to be completely inadequate to permit waste injection.

Citing this lack of knowledge, Piper[288] in 1969 and the National Industrial Pollution Control Council[289] in 1971 proposed a canvass of the United States and immediate research to establish: (1) geological, hydrological, and geochemical factors involved in deep well disposal; (2) areas suitable for injection disposal; (3) a categorization of all wastes based on their suitability for deep well disposal; (4) the legal status of the problem; and (5) effective monitoring procedures for deep wells and disposal areas. These proposed studies were not completed by 1972. Miller[290] also warned of the lack of data on possible groundwater pollution hazards of deep injection wells.

In 1971 Tofflemire and Brezner[291] summarized existing deep well injection practices in the United States, with particular reference to New York State. Site selection, well construction, waste quality, and well-monitoring criteria were explored. Salt water, industrial wastes, and radioactive wastes were the three major types of liquid amenable to deep well disposal. A listing of pertinent current literature was also included.

Rudd[292] in 1972 reviewed Pennsylvania injection wells handling all types of wastes. Drilling and well construction criteria were examined, and the monitoring of system operations, formation pressures, and fluids was discussed. Geologic and hydrologic factors bearing on subsurface disposal of liquid wastes were also detailed by Otton[293] in 1970 for eight major subregions of Maryland.

In 1968 Water Well Journal[294] reported on the construction of an injection well in Middletown, Ohio, to dispose of spent steel mill pickle liquor. The well met rigid state design specifications and, considering the favorable geology, was thought to pose no threat of groundwater contamination.

Only Ohio, West Virginia, and Texas have specific legislation regulating industrial wastewater injection. Cleary and Warner[295,296] presented a monograph on underground wastewater disposal for the Ohio River Basin in 1969. Insight into public policy issues was provided, and administrative and regulatory guidelines were offered to aid in evaluating the location, design, construction, operation, and abandonment of injection wells. Most of the Ohio Valley was considered amenable to waste injection, but it was recommended that only limited quantities of wastes be regarded as eligible for subsurface disposal and that monitoring needs were of great importance.

Two examples of successful drilling and operation of deep injection wells in Indiana were presented by Hundley and Matulis[297] in 1963 and Hartman[298] in 1968. The Newport, Indiana, well disposed of inorganic waste into a sandstone reservoir capped with a sandstone of near zero permeability. In addition, deep well injection at a Midwest Steel Mill plant in Portage, Indiana, was very effective in reducing sludge accumulation. Monthly lime neutralization costs at Portage were reduced by about 80 percent. Design and construction of the well were described.

In 1968 Bergstrom[299,300] studied the criteria for feasibility of industrial waste disposal by injection wells in Illinois, and reviewed the suitability of various geologic formations for disposal. Favorable geohydrologic conditions made disposal by injection wells feasible in much of the southern two-thirds of Illinois. However, exhaustive testing, substantial proof of acceptable site conditions, and incorporation of optimum engineering safeguards were still considered necessary before any well installation could be authorized. The eight basic design policies incorporated into this construction permit system were discussed by Smith[301] in a 1971 survey of subsurface storage and disposal in the state.

Berk[302] in 1971 described the methods used and problems encountered in drilling two deep injection wells in Chicago, Illinois, and Bakersfield, California, for the disposal of industrial liquid wastes. In addition, in analyzing drilling equipment, the study revealed the physical and economic advantages of a combination well casing and injection tubing in the form of saran-lined steel pipe over the standard casing and plastic injection tube design.

An early study by Jones[303] in 1947 of injection wells for the subsurface disposal of Kansas oil field brines emphasized the protection afforded groundwater supplies and the added benefit of repressurizing "played-out" oil fields. It warned of the constant battle with corrosion of brine handling equipment in the use of injection wells.

In 1970 Grubbs, et al.[304] evaluated the geologic and engineering parameters governing the disposal of liquid wastes by deep well injection in Alabama. A study was made of geological and hydrological factors to identify favorable subsurface reservoirs for waste confinement. A design and cost procedure was supplemented by a computer program to provide rapid feasibility studies of proposed sites, and an extensive bibliography was included. Alverson[305] in

1970 reported on a similar evaluation of conditions in Baldwin, Escambia, and Mobile Counties which favored effective deep well injection and outlined criteria to insure against groundwater contamination. In 1971 Tucker[306] reviewed operations of five injection wells over a six-year period, and described effective well design criteria and well monitoring techniques.

Since 1963, a deep injection well system has been operated by Chemstrand Company at Pensacola, Florida, for the disposal of aqueous process wastes from the manufacture of nylon. Batz,[307] Dean,[308] Barraclough,[309] and Goolsby[310] reported on this situation between 1964 and 1971. Without pretreatment, the wastes were injected at low pressures and high rates into the lower limestone of the Floridan aquifer between two thick beds of clay. Dean[308] described the design criteria (especially the casing program), the construction, and the operation of this system in detail. Goolsby[310] employed monitor wells in 1971 to reveal that the waste extended outward about one mile from the two injection wells and that pressure effects extended outward over 25 miles. No detection of wastes above the formation was reported. Lateral travel rates of waste were observed.

In 1969 Garcia-Bengochea and Vernon[311] described deep-well disposal of industrial wastes in the highly saline boulder zone of the Floridan aquifer in Southern Florida at Belle Glade. No trace of groundwater contamination had been found in the overlying aquifer, and the potential of the zone for similar uses, if the hydrogeological conditions were typical of the region, was discussed.

In 1970 Vernon[312] surveyed brine disposal and waste injection wells in Florida. The use of zones of high transmissivities was stressed, and a permit system was described based on proper design and treatment criteria. An extensive monitoring system was considered essential to effective regulation of injection wells. Kaufman[313] in 1973 analyzed data on deep well injection of industrial and municipal effluents and concluded that injected wastes seemed to remain confined in the receiving stratum, at least in Northwest Florida. The crucial need for close monitoring of future hydraulic and geochemical effects was emphasized.

In 1951 and 1953 de Ropp[314] and Henkel[315] followed the development of a waste disposal system at the DuPont Company's adiponitrile plant near Victoria, Texas. Concentrated liquid chemical wastes were injected into a deep well to subterranean sands, and both studies described the untested control techniques employed. Well construction procedures (cementing and anti-corrosive casing) and the entire brine treatment operation (aeration, induced precipitation, filtration, and chlorination) were detailed.

In 1967 Veir[316] and Lockett[317] reported on two deep well disposal systems in Bay City and Odessa, Texas. Both were successful, but the Odessa injection operation proved to be expensive. Geologic and hydrologic sampling programs were discussed, along with details of well construction and equipment and waste pretreatment processes.

McMillion and Maxwell[318] in 1970 reported on field studies conducted in Texas County, Oklahoma, on the pollution potential of the Ogallala aquifer by oil field brine injection. The Glorieta sandstone beneath the Ogallala received the injected brines, and the hydraulic characteristics of the Glorieta were needed to determine the fluid relationship between strata. As a result, a technique was developed for making aquifer tests in brine disposal wells.

Evans[319] and Evans and Bradford[320] in 1966 and 1969 examined the connection between Denver, Colorado, area earthquakes and a deep injection well at the Rocky Mountain Arsenal which was disposing wastes from the manufacture of poison gas. Both reports pointed to the unknown dangers and possible effects of disposal wells, and Evans and Bradford[320] warned that deep injection well techniques offered only temporary safety from the many permanently toxic wastes being injected.

The problems of deep well disposal of industrial and radioactive wastes in Canada were surveyed by McLean[321] in 1968 (with emphasis on Ontario) and van Everdingen and Freeze[322] in 1971. Geological and hydrological features of site selection, well construction and abandonment requirements, waste quality criteria, and monitoring needs were treated.

## RECHARGE WELLS

The University of California Sanitary Engineering Research Laboratory[323,324] detailed a study of pollution travel from direct well recharge. A well field consisting of a recharge well and 23 observation wells penetrating a confined aquifer 100 feet underground was located in Richmond, California. Both fresh water and water degraded with settled sewage were injected at various rates. Chemical, bacteriological, and radiological methods were employed to determine rates of travel of the recharged water. The bacterial pollutants travelled a maximum of 100 feet in the direction of normal groundwater movement even though steep gradients were imposed.

The nature of well clogging was examined, and methods of well redevelopment were studied.

In 1967 Mitchell and Samples[325] reported on an investigation of high rate treatment facilities to polish standard rate activated sludge effluent to make it suitable for use as a water supply for recharge through injection wells. A three-phase investigation (polishing, recharge, monitoring), conducted at the City of Los Angeles Hyperion treatment plant, concluded that either rapid sand filtration with pretreatment or diatomaceous earth filtration could be used to produce water from Hyperion secondary effluent which was acceptable for injection. Cost estimates of the process and suggestions for cooperation with fresh water barrier projects were included in the report.

In 1970 and 1971 Wesner and Baier[326,327] reported on the research of the Orange County Water District in California on wastewater reclamation and subsurface injection. Objectives of the research were to determine: (1) the hydraulic characteristics of the proposed injection barrier system of multi-point injection wells; (2) the long term fate of reclaimed wastewater in the injection system; (3) the feasibility of utilizing wastewater for a barrier; and (4) the chemical composition of blended reclaimed water and deep groundwater. The performance of the system was found to be generally satisfactory, but the persistent odor and taste in the injected reclaimed water was probably the most serious deterrent to utilizing that source for injection in a barrier system.

Since 1966 numerous studies[328 to 338] have been completed on a plan of Nassau County, Long Island, New York, to reclaim water from the effluent of its wastewater treatment plants and inject this treated water through wells into aquifers furnishing most of the public water supplies in the county. The injection would create a hydraulic barrier to prevent salt water intrusion and would allow increased withdrawals from existing water supply wells. A series of artificial recharge experiments has been conducted to determine the feasibility of a proposed fifteen-mile network of barrier injection wells. Ultimately, the plan was to establish numerous water reclamation plants throughout the County for direct injection.

An experimental recharge well was drilled to a depth of 500 feet. It consisted of two fiberglas casings with a 62-foot long stainless steel well screen attached to the bottom of each casing. Hydraulic head changes and water quality were evaluated at several points within the well and filter pack. Geochemical reactions related to the head changes were monitored by means of instruments within each screen.

Various studies of the proposed project have dealt with alternative methods of waste water renovation and comparisons of recharge wells with recharge basins;[329] area withdrawal rates, tertiary treatment costs, and other salt water barrier and advanced waste treatment projects;[332] and recharge water quality standards and methods for achieving them.[333]

In general, experiments since January 1968 have shown that the Magothy aquifer of Long Island can be recharged with reclaimed water through the use of injection wells. Restricted bacterial travel through the aquifer was primarily due to the high filtering efficiency of the fine to medium aquifer sand and to bacterial capture by a filter mat and slime deposits that form around the well during injection. The studies also showed that stringent quality requirements on the injected water were necessary to minimize well clogging and redevelopment.[334 to 338]

## WELL CONSTRUCTION EFFECTS

The connection between groundwater pollution problems and well construction has been explored in recent years. In 1948 Williams[339] summarized a field investigation of contamination of deep water wells in Crawford and Cherokee Counties, Kansas. Three pumping tests revealed the intrusion of highly mineralized saline water (both natural and oil brine) into municipal and industrial wells. The cause was defective well construction; the author recommended proper casing and grouting of new deep wells.

*Johnson National Drillers Journal* [340] reported on a 1955-56 coliform bacteria problem in water pumped from house wells in a suburb of St. Paul, Minnesota. Contaminated surface water had been permitted to drain into the aquifer through a series of drainage wells drilled into limestone. Three possible solutions to the problem were construction of a community water system, drilling of deeper individual wells and casing off of the limestone, and individual chlorination equipment.

A 1963 publication of the U. S. Public Health Service[341] reviewed the problems of installation of chlorinating equipment on individual water systems. The manual was in part intended for use by state and local health authorities, well drillers, and industry groups concerned with the design, construction, and operation of such water supply systems.

Another article in *Johnson Drillers Journal*[342] in 1967 survey-ed problems of control and monitoring of bacterial pollution in well water.  Chlorinated water only was recommended for drilling, along with chlorination of the gravel pack before installation.  Ster-ilization of sampling materials was also stressed.  Standard monitor-ing procedures for discovery of coliform bacteria presence were re-viewed and a new filtration procedure was discussed.

A field study of the water quality in the municipal well of Aron, South Dakota, was detailed by Jorgensen[343] in 1968.  A marked difference appeared between water quality from the muni-cipal well and nearby wells tapping the same aquifer.  An aquifer test and analyses of water samples showed the anomaly to be caused by leakage from a nearby abandoned well tapping another aquifer.

A 1971 Ham[344] surveyed the general problem of groundwater pollution through wells and presented diagrams of various avenues of pollution via wells.  A list of eleven items dealing with statutory and administrative control measures was included.

Also, in 1971 Jones[345] concluded that since most groundwater aquifers have multibarrier natural defenses, the path of entrance for most contamination was the well itself.  Only rarely did circum-stances economically justify substitution of disinfection techniques for adequate protection of the groundwater source.  Poorly con-structed and abandoned wells served as unauthorized and uncon-trolled recharge points and had a degrading effect on groundwater quality.  In order to achieve water quality improvement, well drill-ers, water conditioning dealers, and county and state health depart-ments had to accept as the ideal goal the exclusion of water of un-desirable quality from sources of groundwater supply.

# CHAPTER VIII

# SALT WATER AND SURFACE WATER

## SALT WATER INTRUSION

Salt water intrusion is characterized by the movement of saline water into a freshwater aquifer. Almost all of the literature in recent years has been devoted to sea water intrusion of coastal aquifers, while the upward movement of brackish or saline waters from connate sources in inland aquifers has been relatively neglected.

Parker[346] in 1955 presented a general discussion of sea water intrusion together with examples from various locations in the United States. Five years later Todd[347] surveyed the coastal intrusion situation in the United States, prepared a map of intrusion locations, and mentioned control efforts in Los Angeles, California. In 1968 a symposium on the subject was held at Louisiana State University;[348] papers covered salt water encroachment into aquifers in Florida, New York, and California, management of aquifers, encroachment control, hydrogeology, and legal aspects of encroachment.

The most recent report on U.S. saline intrusion was prepared in 1969 by a Task Group of the American Society of Civil Engineers.[349] General mechanisms responsible for intrusion include (a) reversal or reduction of groundwater gradients, (b) accidental or inadvertent destruction of natural barriers that prevented movement of salt waters, and (c) accidental or inadvertent disposal of waste saline water. The occurrence of saline water was described in terms of geologic and hydrologic conditions. The extent of intrusion was listed with 68 examples from inland and coastal aquifers throughout the United States. Major control efforts underway in some areas were also mentioned.

A field investigation of sea water intrusion on the southern coast of Long Island, conducted in 1958-61,[350] found that saline groundwater occurred both in permeable deposits and clay deposits. Chloride contents were only slightly less than that of sea water, and a broad transition zone was present. Movement of the saline water inland and downward was very slow; there was no danger of public supply wells being affected within the next two decades. Data were obtained by test drilling, extraction of water from cores, electric logging, water sampling, and water level measurements. New concepts of environmental-water head proved useful in defining hydraulic gradients and flow rates of groundwater having a variable density. A suggested control mechanism was artificial recharge of imported water to create hydraulic barriers.

A follow-up study on the same Long Island area by Cohen and Kimmel[351] in 1969 covered the period 1960-69. Landward movement of a deep wedge of salty groundwater in the area was minimal; significant changes in chloride content were noted in only 3 of 30 outpost wells. These increases resulted from local heavy pumping near the zone of diffusion. No increases were noted in the underlying Lloyd aquifer, except where leaky casings permitted downward flow of saline water.

The status of sea water intrusion in Delaware aquifers was reported by Woodruff[352] in 1969. Brackish water was present in nearly all aquifers, but the depth and location of the fresh-salt interface varied with each aquifer. Intrusion in shallow water table aquifers was spotty. It was stated that properly located monitor wells were necessary for detection of future chloride movements. To control intrusion heavy pumping should be avoided near brackish water areas.

Three reports in 1964 documented the salt water intrusion situation in the Savannah area of Georgia and South Carolina.[353, 354, 355] Fresh water in the principal artesian aquifer was being contaminated by two sources—in the upper zones by sea water intrusion, and in the lower zones by incompletely flushed water of Pleistocene Age. The rate of salt water movement was slow, so that intrusion was not expected to reach Savannah for nearly 100 years. It was recommended that additional deep observation wells should be drilled to define the intrusion pattern. To control intrusion pumping should be regulated and rearranged over a wide area; furthermore, surface water should be developed as a supplemental source.

Siple[356] in 1965 described a detailed investigation of sea water intrusion in the limestone along the South Carolina coast. Geological, hydrological, and geochemical methods were employed. Carbon-14 measurements of groundwater at Hilton Head, South Carolina, were used by Back, et al.[357] to define trends and areas of sea water intrusion there.

Along the coast of Georgia gamma-radiation logs were used by Wait[358] to identify phosphate zones. It was concluded that mining of one phosphate zone would breach the confining layer above the principal artesian aquifer and allow sea water intrusion to occur. In a study of water-level declines in the principal artesian aquifer in Glynn County, Georgia, by Gregg,[359] a head imbalance was found between the aquifer and an underlying brackish-water zone. Leaks of brackish water upward into the aquifer occur through breaks in the confining layer. A relief well drilled into the brackish-water zone was pumped at 3,000 gpm to create a hydrostatic equilibrium with the aquifer. The pumping was apparently successful as successive water samples from the aquifer showed a decrease in chloride content. More relief wells may ultimately be needed to control the intrusion.

Wait and Callahan[360] in 1965 made a survey of sea water intrusion along the southern U.S. Atlantic Coast. Intrusion was attributed to lateral and vertical encroachment due to adverse hydraulic gradients, percolation, inundation of fresh water lenses in storms, overpumping, connate saline water, and drainage canals. Detailed descriptions of fresh-salt water conditions in five areas, extending from North Carolina to Miami, Florida, were presented.

Florida, with its long and heavily populated coastline, has been severely affected by sea water intrusion in its coastal aquifers. An early comprehensive analysis of the situation was prepared by Black, et al.[361] in 1953. The history, extent, theoretical basis, factors responsible, and examples were described. High chloride contents of groundwater resulted both from connate waters and from sea water. Intrusion was a direct result of large withdrawals for cities, agriculture, and industries, which lowered piezometric heads. In addition, excessive drainage in areas of low piezometric head was a contributing factor. Chloride was the most reliable index of salt water encroachment.

In the Miami area of Florida, sea water intrusion was caused by inadequately controlled tidal drainage canals which reduced fresh groundwater heads and provided access paths for sea water to move inland during dry periods. In a study of the problem Klein[362] reported that properly placed water control structures, such as tide gates, retarded and reversed the encroachment. Similarly, Kohout[363] described how construction of a storm sewer in Miami caused localized sea water intrusion. To control the problem a sheet pile dam was built at the sewer outlet; observation wells in the vicinity showed salinity variations occurring before and after construction of the dam.

Field measurements on the Biscayne aquifer of the Miami area in Florida were made by Kohout and Klein.[364]   With heavy rainfall a pulse recharge of fresh water occurred which contributed to the full thickness of the aquifer, whereas in other periods fresh water flowed outward through only about one-third of the aquifer thickness.

An unusual hydrologic phenomenon involving sea water intrusion was reported in 1969 at Tarpon Springs, Florida, by Stringfield and LeGrand.[365]   Here deep vertical sinkholes serve as openings into a carbonate aquifer.   Sea water entering the sinkholes caused fluctuations in the dynamic equilibrium between fresh water and salt water in the aquifer.   As a result the flow of salt water from the spring (a sinkhole) to a lake two miles away was sometimes reversed.

In 1972 Sproul, et al.[366] described how upward leakage of saline water from an artesian aquifer 1500 feet deep raised chloride contents and temperatures of groundwater in the Lower Hawthorn aquifer of Lee County, Florida.   The water moved through wells or along a fault or fracture system and spread laterally over a 2.5 square mile area.   Saline water then moved upward into the Upper Hawthorn aquifer, which is the principal source of public water supplies, through wells connecting the aquifers and by infiltration of water discharged at land surface from wells tapping the Lower Hawthorn aquifer.   To prevent upward movement of saline water, cement plugs should be set in well bores to separate the aquifers.

*Ground Water Age*[367] in 1973 discussed the use of 3000-foot injection wells to prevent salt water intrusion at St. Petersburg, Florida.   Secondary treated domestic wastewater, surface runoff, or shallow aquifer seepage water could be injected to create an artificial shallow aquifer mound to retard intrusion of sea water.

Southwestern Louisiana has been affected by sea water intrusion, according to several investigators.   In the Baton Rouge area three reports have concerned intrusion.[368,369,370]   Saltwater encroachment had advanced northward due to heavy pumping.   Previous estimates of salt water reaching Baton Rouge in 5 to 10 years, however, no longer applied; a fault was discovered which acts as a barrier to protect pumping centers from the pollution.

Removal of the small amount of salt water north of the fault will result in natural replenishment with fresh water.

The feasibility of a scavenger-well system to solve vertical salt water encroachment was studied by Long[371] at Gonzales, Louisiana, in 1965. An existing well screen was divided with a packer so that fresh water was pumped from the upper section and saline water from the lower. The system was effective, but it was believed that two separate wells would be better. The chloride content of water from a nearby supply well varied with the ratio of pumping rates of the scavenger well.

Harder, et al.[372] prepared a detailed report on intrusion in Southwestern Louisiana. Heavy groundwater pumping lowered water levels and caused sea water encroachment from the south and east toward concentrated withdrawal areas. The advancement was occurring at 20 to 300 feet per year. There was no serious threat to water users, but some deep wells were abandoned. The saline water contained significant quantities of bromide and iodide. To minimize the intrusion new well fields should be located away from centers of heavy withdrawals.

In Texas a study in the 1950's was concerned with sea water intrusion in the Houston area.[373,374] The encroachment was attributed to a heavy concentration of pumping which reversed the natural hydraulic gradient. Less probable potential sources included upward movement of salt water from below, vertical movement around salt domes or along faults, downward seepage from surface sources, and pollution through leaking wells. The rate of advance was very slow; the closest salt water was probably 5 miles from a pumping center in the deepest sands. Strategically placed observation wells were needed in addition to 70 wells which were periodically sampled over the last 20 years.

The California coastline contains numerous localities affected by sea water intrusion. One of the earliest detailed studies was that of the Santa Ana-Long Beach area by Piper, et al.[375] in 1953. The potential sources of salt-water pollution were defined (the ocean, connate water, oil field waste fluids, and industrial fluid wastes in streams), the lateral extent of intrusion was described, and the tendency to further reductions in groundwater quality was analyzed. Base-exchange substitution of calcium and magnesium for sodium was common, making it usually impossible to discriminate among various salt-water sources. Iodide or borate were seldom reported, but it was suggested that these, together

with barium, should be reported in future analyses to aid in iden-
tifying pollution sources.

Beginning in 1953 the California Department of Water Resources studied sea water intrusion with particular emphasis on
Los Angeles County.[376, 377, 378]  The entire intrusion situation in
California was reviewed; in some areas salt water had moved up
to 4½ miles inland, and continued pumping would allow further
encroachment.[376]  Control methods considered included (a) reduction in pumping, rearrangement in the areal pattern of pumping, or both, (b) direct recharge, (c) maintenance of a fresh water
ridge above sea level along the coast, (d) construction of an artificial subsurface barrier, and (e) development of a pumping trough
along the coast.  It was emphasized that suitable standards for
well construction and abandonment should be established and en-
forced; existing laws were inadequate.

A second phase of the study[377] was a field investigation of
the feasibility of controlling intrusion.  A 5,000-foot line of recharge wells was constructed in a confined coastal aquifer at Manhattan Beach, California.  Injection of treated imported fresh water was found to reverse the landward hydraulic gradient and to
prevent further intrusion.  The recharged water will aid in replenishing the aquifer, and formerly polluted portions of the aquifer can be reclaimed.  Detailed information was presented on the
quality of recharge water, effect of aquifer transmissibility, chlorination of recharge water, geology, hydrology, maintenance problems, and project costs.

The third phase of the study[378] included a literature review,
theoretical analyses of the mechanics of sea water intrusion and
its control, and a laboratory study on the feasibility of subsurface
barriers.

After the extensive Los Angeles study of intrusion, the California Department of Water Resources began a continuing series
of local investigations at other localities along the California coast.
As part of the San Dieguito River investigation in 1959, sea water intrusion was found along the coastal margin because of over-
draft on the limited aquifer.[379]

In 1960 a report was issued on intrusion in Southern Alameda County.[380]  The intrusion was caused by sea water entering
through aquifer gravels exposed to tidal currents of San Francisco
Bay, by aquiclude leakage to the underlying aquifer, and by sea
water entering through abandoned, defective, and improperly con-

structed wells. The need for suitable standards for well construction and sealing of abandoned wells was emphasized. The search for and sealing of problem wells was to be continued.

A report on intrusion in the Oxnard Plain of Ventura County, California, appeared in 1965.[381] Here salt water in the upper aquifer had moved 2 miles inland, was advancing at about 1,000 feet per year, and had rendered 44 wells useless. There was no evidence of intrusion in the underlying aquifers; however, degraded water may move downward through wells or natural breaks in confining beds.

Later a test of the pumping trough method of intrusion control was made in the same Oxnard Basin.[382] Five experimental extraction wells operating for two years reduced the areal extent of degraded water by 15-20 percent. The barrier was shown to be technically feasible; however, more study was needed to determine well spacings and pumping rates. A further report[383] in the same area, released in 1971, involved field, laboratory, and theoretical analyses of the influence of aquitards on sea water intrusion.

In 1966 a study on the Santa Ana Gap in Orange County, California, was published.[384] Sea water intrusion due to over-pumping extended 4 miles inland; other sources of degradation included improperly discharged oil field brines and upwelling of connate waters. Salinity control barriers considered were static, pumping trough, injection ridge, and combination of pumping trough and injection ridge. The combination barrier was regarded as most practical to avoid problems of waterlogging and subsidence. Treated wastewater was the most economical source of continuously available water for the injection phase of the barrier.

The nearby Bolsa-Sunset area in Orange County, California, was the subject of a 1968 report.[385] The Newport-Inglewood fault, which approximately parallels the coast, forms a hydraulic barrier across the area except in late Recent deposits. Pumping in 1945-57 caused sea water intrusion through permeable portions of the fault and through aquiclude discontinuities. Artificial recharge of the forebay caused a recovery of piezometric levels during 1959-65. Intrusion and brine wedges have retarded or have become stabilized since 1961.

In 1970 intrusion in the Pismo-Guadalupe area of San Luis Obispo County, California, was reported.[386] Increasing chloride concentrations in wells penetrating shallow aquifers were traced to the natural salinity of the geologic environment, salt concentration by evapotranspiration, and downward percolation of sea water entering tidal channels at times of extremely high tide. There was evidence of intrusion in three deeper aquifers but no danger to water supplies. Data were obtained from a grid of monitoring wells sampled semiannually and annually.

The most recent intrusion report by the California Department of Water Resources appeared in 1972 on the Morro Bay area, also in San Luis Obsipo County, California.[387] Several wells were abandoned due to intrusion, primarily due to lowered water levels from intensive pumping. The extent of intrusion had been controlled by seaward underflow during periods of low pumpage. Heavy pumpage from lower aquifers could induce downward migration of sea water through low permeability layers.

Other California sea water intrusion publications include a paper[388] in 1964 on the proposed use of reclaimed water for injection to control intrusion in Orange County, two reports[389, 390] on the West Coast Basin barrier project in Los Angeles County, a report on the Alamitos barrier project[391] in Los Angeles and Orange Counties, and an analysis of economic and legal implications of intrusion in Salinas Valley.[392]

The problem of sea water invading aquifers in Hawaii has been recognized since early in the century. Lau[393] in 1967 studied the equilibrium conditions existing between fresh and salt waters, particularly as they apply to Oahu conditions. Todd and Meyer[394] analyzed aquifer conditions in Honolulu, related pumping rates to increases in chloride content, found that vertical displacement of salinity near pumping wells varied inversely with salt concentration, and computed natural recharge rates as a guide to maximum pumping rates so as to avoid increased intrusion.

Intrusion was described at Summerside, Prince Edward Island, Canada, by Tremblay, et al.[395] in 1973. Salt water movement occurred in an upper zone due to overpumping and in a lower zone by intermittent pumping which caused a thickening of the diffusion zone. Control measures suggested included limiting the depth of wells, avoiding intermittent pumping, and pumping salt water from the lower zone.

Research on salt water intrusion has been active in recent years; consequently, several important contributions to a better understanding of intrusion have appeared. A series of field and theoretical investigations by the U.S. Geological Survey were reported by Cooper, et al.[396] in 1964. The five parts included a concept of circulation of sea water due to dispersion of salts in the zone of diffusion, results of a field investigation of the phenomenon at Miami, Florida, two mathematical solutions for the position of a sharp interface that would occur without diffusion, and the effects of dispersion.

The effectiveness of a fresh-water canal to act as a barrier to salt water intrusion was treated in two papers in 1967.[397, 398] Numerical results were presented for a canal paralleling the sea. Solutions showed that fresh-water flow from a canal acts as a dam, forcing the fresh-salt water interface to a lower elevation. Use of a canal for reclamation of salt-water intruded deltas and marshes for agricultural purposes was discussed.

Laboratory studies of a method for flushing water from aquifers into subsurface drains by applying fresh water to the surface were described by Carlson and Enger[399] in 1969. With the drains in operation, a stable interface was formed. Reducing the drain spacing reduced the amount of salt water removed. A numerical technique for calculating the transient position of a salt-water front in a coastal aquifer was reported by Pinder and Cooper [400] in 1970. The method of characteristics was used to solve the solute transport equation, and the alternating direction iterative procedure led to solutions of the groundwater flow equation for two-dimensional flow in nonhomogeneous aquifers with irregular geometry.

During the period 1970-72 Kashef[401, 402, 403] published three papers on sea water intrusion. The first described viscous flow models and their utility as well as other types of hydraulic models for intrusion problems. The second stressed the roles of groundwater management and basic research in solving intrusion problems, and the third provided a historical review of different approaches to the analysis of intrusion.

Peek[404] in 1969 studied the effects of large-scale phosphate mining in Beaufort County, North Carolina. Field data showed that overpumping for the mining led to salt water encroachment from three sources. Control measures advocated included reduction in pumping, new well spacings, withdrawal of brackish water, and artificial recharge.

In a study of Chaves County, New Mexico, Hennighausen[405] found that salt water intrusion, presumably from connate sources, was occurring both laterally and vertically by overpumping which was reducing the artesian head. To control intrusion pumping would have to be reduced.

## SURFACE WATER

Brine effluent discharged from a municipal water treatment plant in Lancaster, Ohio, into a river above the municipal water supply aquifer was the subject of a 1962-64 field investigation by Norris.[406]  The resulting chloride concentrations in the groundwater remained low ($\approx$38 ppm) because discharged muck, organic debris, and iron deposits had resulted in a low streambed permeability, allowing only a low rate of direct surface water infiltration.  Some of the wells at the 100-foot depth levels revealed organic contamination in the form of red slime deposits and strong sulfide odors.  Suggested control methods included halting discharge of brine wastes, restoring streambed permeability by dredging, construction of an off-channel recharge pond, and the widening of levees.

In 1963 Klaer[407] discussed the process of induced infiltration of water from surface streams to aquifers.  In such cases, natural sand and gravel deposits served as large natural filter beds, effectively removing or reducing turbidity, organic material, and pathogenic bacteria.  The paper analyzed the general processes by which such removal was accomplished, as well as the significance of certain changes in chemical characteristics of the water as it passed from a surface source to an underground point of collection.

Preul and Popat[408] in 1967 presented predictive mathematical models to determine the quantity and quality of recharge from the Great Miami River in Southwest Ohio to two collector wells in an adjacent aquifer.  The concentration of pollutants was determined as the sum of concentrations introduced by the convective flux between the river and the wells.  Calculated pollutant concentrations were tabulated and compared with measured concentrations.  It was concluded that a high percentage of the water recharging the two collector wells originated from the river, and that the adsorption and ion exchange capabilities of the aquifer were nearly exhausted.

In 1970 Randall[409] reported on the presence of coliform bacteria in a municipal well in Binghamton, New York.  The well produced from beds of coarse sand and gravel 10 to 80 feet below land surface, and had had 19 years of trouble-free operation until 1964.  The coliform bacteria traveled at least 180 feet to the well from a reach of the Susquehanna River polluted by sewage.  The cause seemed to be the excavation of the riverbed in an area already geologically favorable for induced infiltration.

# CHAPTER IX

# POLLUTANTS AND EFFECTS

## GENERAL POLLUTION STUDIES

A large number of references pertaining to groundwater pollution generally or to pollution in an area or region from multiple causes have appeared in the literature within the last 20 years. This subsection briefly reviews these contributions.

The American Water Works Association has been concerned with underground waste disposal and contamination of groundwaters since 1932. A task group on this subject has made periodic assessments of the problem. In 1952[410] it reported that the problem was of rather wide distribution, that it varied in severity from region to region, and that much more data would need to be assembled. By 1957[411] it described underground pollution as a national problem with many variations; tabulated sources, contaminants, statutory controls of states, and pollution travel distances; and suggested that pollution trends were related to population and industrial activity. Finally, in 1960 two reports[412,413] appeared. The first stated that industries and legislative bodies were becoming increasingly aware of the problem, that much work and many precautions were necessary to insure satisfactory conditions, and that flexible standards should be developed for guides in evaluating disposal techniques. The second was a survey of all types of subsurface pollution with lists of sources and statutory controls for all states. Special comments were addressed to synthetic detergents, well regulation, and sewage lagoons.

The World Health Organization[414] in 1957 discussed the danger of groundwater pollution by disposal of wastes into wells or pits. It was mentioned that such practices should be restricted to where the receiving aquifer is unfit for other uses and there is no possibility that the waste will move into other water-bearing strata. Once groundwater is polluted, the chances of effective remedial programs are remote. It was suggested that each country should proceed toward the adoption of a national water policy.

Rorabaugh[415] in 1960 discussed problems of waste disposal and groundwater quality. He pointed out that wastes reach aquifers both intentionally and unintentionally. Analysis of pollution underground can only be approximated because of problems of head differences between aquifers, differences in temperature and density, and diffusion. In the following year Bolton[416] listed sources of groundwater contamination. A frequent cause is the storage of wastes on ground surface. A national master plan for preventing pollution of water resources was proposed.

An analytic survey of the salt balance problem in groundwater was reported in 1963 by Meron and Ludwig.[417] Salt increases result from importations, irrigation, fertilizers, municipal and industrial use, oil brines, and sea water intrusion. Detailed quantitative data were presented for groundwater basins in Los Angeles and Orange Counties, California. Control methods discussed included evaporation of highly saline wastes, transport of wastes to the sea, control of waste-producing operations, importations of high quality water, injection wells, and demineralization.

In a series of papers LeGrand[418 to 421] discussed the problems and management of groundwater pollution. He stressed the need for understanding pollution as a first step to control. Management of the problem requires that hydrogeologic environments be classified along lines of the interdependence of factors such as permeability, sorption, hydraulic gradient, position of water table, and distance from contamination source. Long-range plans for urban areas and suburbs should include recognition of the deterioration of groundwater quality made possible by waste disposal practices, artificial recharge, accidents, and the presence underground of salt water. Technically trained personnel capable of determining the best use of land for water supply and waste disposal are rarely part of administrative procedures. Specific problems are more often treated than long-range planning.

Finally, in 1965 LeGrand[422] evaluated patterns of contaminated zones of groundwater. A wide variety of pollution plumes from point sources were diagramed and explained. Attenuation effects due to dilution, decay, and sorption were described for various types of pollutants.

A survey by Rainwater[423] in 1965 covered all types of groundwater pollution. He reported that 36 states had anti-groundwater pollution laws. Control mechanisms described included skimming wells, specially constructed and sealed wells, brine pumping, and isolation of brine springs.

In 1966 Oltman[424] surveyed research underway on ground-water pollution under the Water Resources Research Act of 1964. Projects included hazards of water supply and sewage disposal systems, sanitary landfills, nutrients in sewage effluents, viral pollution, leaching effects, and algal growth.

The hazards to groundwater from addition of manmade wastes were reviewed by McGauhey[425] in 1968. Experiments showed that bacteria and viruses generally do not move more than a few hundred feet in soil. Dissolved products of biodegradation of wastes move freely and increase salt concentrations of groundwater. Chemicals added include metal ions, phenols, tars, brines, and exotic organic compounds. Agriculture may enrich water with dissolved soil constituents, nutrients, and pesticide residues. Leaching from landfills may involve chemicals, oils, iron, and earth constituents. The most serious contamination hazard in groundwater is the increase of dissolved solids to levels unfavorable for beneficial uses.

The pollution of groundwater viewed from a legal standpoint was discussed in 1969.[426] Waste disposal and salinity were described as the two major sources of pollution. The problem of tracing pollution underground makes it difficult to set disposal standards. Natural purification processes should not be relied on for protection from contamination. Archaic legal doctrines hamper the effectiveness of control programs. A reevaluation of the current water-rights doctrine is necessary, and workable means to control aquifer withdrawals to prevent intrusion must be developed.

In 1970 the editors of *Water Well Journal* devoted an entire issue to the subject of groundwater pollution.[427] The 12 essays cover the occurrence, use, and protection of groundwater resources; classification of pollutants; sources of pollution; purification of groundwater; and regulatory controls by government agencies.

A survey of municipal water supply systems led to a report in 1970[428] on biological problems encountered in water supplies. Iron bacteria in wells was the most common groundwater problem reported.

Evidence of the increasing attention being focused on groundwater pollution was the fact that eight papers on the general subject appeared in 1972. Kazmann[429] explained that aquifers can be used for many purposes other than as water supply sources and that these activities can be safely accomplished if provisions for monitoring are made to detect their impact on water supplies before deleterious effects occur. Pettyjohn[430] described how pollution problems have been caused by inadequate waste disposal in the past, including examples of a livery stable, a gas works, a burial ground, an outdoor privy, and cavernous limestone. Callahan [431] reported on the role of the U.S. Geological Survey in studying groundwater pollution and on the new program of the Survey to study the fate of wastes deliberately or accidentally placed in the subsurface. Wood[432] reviewed the causes of groundwater pollution, the recent innovation of injection wells, and the need for monitoring; he emphasized that the most satisfactory cure was prevention. Lewicke[433] surveyed the various problems of subsurface pollution and suggested the development of a national policy for groundwater protection. Hughes and Cartwright[434] reviewed research on the contamination effects of landfills and septic tanks and suggested that controls should involve geologic site selection and proper design of disposal systems. In a literature review of nonpoint rural sources of water pollution, Lin[435] summarized information on various agricultural activities that affect groundwater quality. Finally, Walker[436] discussed mechanisms of toxic chemical disposal; the need for factual information, corrective measures, and governmental regulations; and the fact that economical and readily accessible measuring equipment for all toxic chemicals have not been developed.

Turning now to groundwater pollution occurrences in various states, two reports by Motts and Saines[437,438] analyzed the causes and trends of groundwater pollution in Massachusetts. From 1850 to 1966 a 4- to 10-fold increase of chlorides occurred throughout the state, accompanied by an accelerated increase from 1954 to 1966. If this trend continues, numerous aquifers could become unusable. Fecal pollution had occurred locally where rocks had fracture porosity.

A report on groundwater quality in a portion of Suffolk County, Long Island, New York, in 1970 by Perlmutter and Guerrara[439] showed that detergents (MBAS) were widely distributed through the aquifers, had increased during the period 1961-66, and were largely stabilized thereafter until 1968. Chloride concentrations had an upward trend, presumably due to sewage effluent and deicing salts. Nitrates were slightly higher. Sewer systems should improve the quality of groundwater.

A field study of groundwater quality near Raleigh, North Carolina, over the period 1962-65 was summarized by Chemerys. [440] Detergents were found to be insignificant, probably due to

the thick sandy clay soil which limited water movement from septic tanks to wells.    New biodegradable detergents should prevent
future detergent problems.    Water from one-sixth of the 60 wells
sampled showed high chloride and nitrate contents.

In 1971 Parizek[441] discussed the influences of land use on
groundwater quality in carbonate terranes of Southeastern Pennsylvania.    Included were comments on agricultural activities, solid
and liquid waste disposal, and gasoline leaks.    In West Virginia
Wilmoth[442] described saline groundwater problems at five sites.
Salt concentrations before, during, and after subsurface industrial
activities and road salt piling were detailed.    After stoppage of
contaminating sources, chloride concentrations returned to former
levels within about 10 years.

For Michigan an early discussion of sources and problems of
subsurface pollution was presented by Billings.[443]    Later in 1963
Deutsch[444] prepared a definitive report on the sources and causes
of groundwater contamination within the state.    In spite of the
numerous sources, he found that the total volume of adversely
affected groundwater was only a small part of the total resource.
Pollution due to oil and gas production had been virtually eliminated by government supervision of construction and operation of
wells.    Sewers, treatment plants, and abandonment of septic tanks
were necessary to limit pollution.    More data and legal controls
were required to regulate saline intrusion of aquifers.    Recently
Burt[445] described the interrelationships between an industrial
plant in Michigan and a shallow aquifer.    The effects of industrial and human wastes on the aquifer were presented together
with the corrective actions taken to protect the groundwater resource.

Sources of groundwater pollution in Indiana were surveyed
by Jordan[446] in 1962.    In Kentucky the problem of waste disposal into karst aquifers was reviewed by George[447] in 1973.    To
control the situation he suggested education of the public, planning, discontinuance of dumping, and government supervision of
sanitary landfills.

Hackett[448] in 1965 discussed the sources of pollution in
urbanized Northeastern Illinois.    Regional planning, development
of hydrogeologic criteria, and establishment of engineering specifications were advocated to cope with groundwater contamination.
Four years later Walker[449] summarized Illinois groundwater pollution and precautionary measures.    The contamination hazards of
aquifers in the state were shown by maps which can be used to

guide planning of locations of wells, oil wells, and garbage disposal sites. Hazards were highest in highly permeable glacial outwash and in outcropping cavernous carbonate rocks, and lowest in buried consolidated fine sandstones.

In Minnesota Wikre[450] presented a comprehensive paper in 1973 on the groundwater pollution problems of the state and suggested the need for source regulation and in some instances of pumping polluted groundwater to waste. Williams[451] described the subsurface pollution situation for Missouri. Pollution hazards were widespread in Southern Missouri where permeable soils and cavernous bedrock exist, whereas they are more localized in Northern Missouri. The problems can be eliminated by an awareness of geology, by planning, by adequate funding, and by authority to follow planning. In Kansas Foley and Latta[452] in 1966 surveyed the saline contamination from various industrial sources. Regulations for brine disposal, well construction, and well abandonment were necessary for control.

Four studies have reported on underground contamination in Colorado. Gahr[453] in 1961 described gasoline and sewage, chemical, and industrial waste pollution near Denver. To control these problems pumping of gasoline, lining disposal pits, constructing deep injection wells, and treating all wastes were proposed. Later in the same area Page and Wayman[454] conducted a field and laboratory study of ABS, bacteria, and dissolved solids from sewage sources. ABS was not significantly reduced by most soils bacteria removal was high, and dissolved solids were unchanged.

Aquifer pollution near Windsor, Colorado, was analyzed in 1966 by White and Sunada.[455] The several pollution sources were identified in the area, and an overall study of the basin was made using a mass balance approach. Pollution sources contributed very little contamination; the primary cause of the increase in total dissolved solids, estimated at 173 ppm/yr, was the high evapotranspiration rate in relation to the groundwater and surface water outflows. The final Colorado study was a field investigation of groundwater pollution by oilfield brine, ABS, and nitrate in the Middle and Lower South Platte River Basin.[456]

For three areas in Utah, Handy, et al.[457] described increases in total dissolved solids in the groundwater. Responsible factors included application of irrigation water, disintegration of crop debris, fertilizers, transpiration, reuse of water, and subsurface saline water migration caused by pumping. In Oregon, Brown[458] reported in 1963 on deterioration of groundwater quality in part of Portland. Causes were overpumping, which had produced an upward movement of saline water into the basalt aquifer, and artificial recharge of industrial cooling water. Need for more studies and for a comprehensive control plan was stated.

In California several studies have been made by the State Department of Water Resources on changes in groundwater quality at various locations. A field investigation in Ventura County [459] attributed degradation to infiltration of surface water, sea water intrusion, interaquifer flow, and movement through wells. On the west side of the Lower San Joaquin Valley salinity increases resulted from upward movement of connate water and from lack of drainage facilities on irrigated lands.[460] In the Mojave River Valley[461] users of groundwater complained of tastes, odors, and foaming from sewage plant effluents and railroad wastes. In the Lompoc area of the Santa Ynez River Valley[462] increasing sulfate and chloride concentrations were noted; they resulted from irrigation return flows and/or upward intrusion of connate water.

A study of the San Diego Region in California[463] revealed groundwater quality impairment from sea water and connate water intrusion, wastewater disposal practices, and irrigation return flows. The control needed to improve quality was flushing of the aquifers with imported, desalinized, or reclaimed water. An analysis in the Santa Clara River Valley[464] showed that groundwater was being degraded by agricultural practices and by urbanization. Because the natural quality was already marginal, a management program to control pollution was recommended.

Finally, a recent paper by Orlob and Dendy[465] applied a systems approach to water quality management in the Santa Ana Basin of California. Beginning with a recognition of sea water intrusion and wastewater return flows as primary pollution sources, a management plan was described incorporating quality standards and costs.

## BACTERIA AND VIRUSES

Many waterborne outbreaks of viral diseases have involved small well-water supplies contaminated by effluents from subsurface wastewater disposal systems. Recent studies have investigated the extent to which soil acts as an agent in the transmission of waterborne viruses.

A 1956-1957 study by Ritter and Hausler[466] of coliform and enterococci organisms in 13 rural wells near Lawrence, Kansas, showed wide variation in MPN values per month for all wells. In addition, enterococci were found more frequently than coliforms in wells of good location and construction, suggesting that one sampling is not sufficient to evaluate the sanitary quality of well water.

In 1968 Drewry and Eliassen[467] reported on experiments showing that virus retention by soils is an adsorption process characterized by linear adsorption isotherms. Water containing radioisotope-tagged viruses was passed through columns of various soils, and the adsorption was dependent on the pH of the soil-water system, with maximum adsorption at 7.0-7.5 pH. The adsorption by some soils was enchanced by the cation concentration in the liquid, but in general the ability of a soil to adsorb viruses cannot be judged by the normally measured soil characteristics. The authors concluded that virus movement through saturated soils should present little hazard to groundwater supplies, provided soil strata are continuous and the usual public health practices of separation of wells and disposal systems are followed.

Results of a similar study by Drewry[468] also showed that virus adsorption by soils is greatly affected by the pH, ionic strength, soil-water ratio of the soil water system, and various soil properties. Virus movement through soils seemed to be unaffected by the degree of pollution of the water. Carlson,[469] too, has conducted research on the adsorption of two types of viruses, measuring various clay and chemical effects.

Laboratory experiments have been conducted on the migration of viruses in percolating water through three Oahu, Hawaii, soil types. Wahiawa, Lahaina, and Tantalus cinder soils were subjected to concentrations of bacteriophage T4 by Tanimoto, et al.[470] and of poliovirus type 2 by Hori, et al.[471] and studied for their ability to adsorb the viruses. The studies simulated the action of a cesspool leaching into the ground. The Wahiawa and Lahaina soils adsorbed 100 percent of the bacteriophage at depths greater than 2½ inches and at a concentration of 2.5 x $10^6$ virus per ml of feed solution. Likewise, the removal of the poliovirus was 97 percent at soil thicknesses of 6½, 2½, and 1½ inches and virus concentration of 150,000 PFU/ml of feed solution. The Tantalus, however, was ineffective in retaining the bacteriophage at thicknesses up to 15 inches, and removed only 22 to 61 percent of the poliovirus at the three depths studied.

Coliform bacteria from unspecified sources have been detected within two miles of Portales, New Mexico, in a recent preliminary examination of the Ogallala aquifer in that vicinity by Bigbee and Taylor.[472] The coliform organisms were seen as bacterial indicators of fecal pollution, and their presence in the water table indicated pollution by aquifer recharge.

Poliovirus and coliform organisms were found in a well water supply in Monroe County, Michigan, in 1970.[473] The well penetrated a limestone formation beneath a shallow layer of glacial drift. Pollution could have entered through surface openings of the well or from a septic tank 100 feet away. Corrective measures included construction of a new well and the banning of shallow disposal wells in the county.

## DETERGENTS

In 1960 Delutz[474] summarized the results of a study of synthetic detergents in the well waters of Portsmouth, Rhode Island, and in wells across the state. The presence of synthetic detergents was the result of leaching, with septic tanks and absorption fields effecting only partial removal. Groundwater samples from 72 wells contained detergents; furthermore, a high correlation was found between degree of chemical and bacteriological contamination and proximity to sewage disposal units. Thorough analyses of the Portsmouth wells showed 24 of 25 with detergents, indicating seepage from a sewage disposal field. Lot sizes of at least two acres were recommended where no public water facilities existed, and a minimum distance of 100 feet between any well and any sewage disposal unit was considered a necessity.

In 1962 Perlmutter, et al.[475] conducted a field investigation of detergent contamination of groundwater in the South Farmingdale area, Long Island, New York. Concentrations up to 32 ppm of alkylbenzene sulfonate (ABS) were found in the upper twenty feet of the water table aquifer by means of tentative methods using methylene blue dye, chloroform extractions, and colorimetric comparisons. Most of the remaining groundwater contained less than 1 ppm ABS and did not foam. The sources of contamination were hundreds of randomly distributed cesspools and septic tanks in the area. As of 1964, the movement of the contaminated groundwater was lateral, but it was feared overpumping might induce downward movement to public supply well depths. The only practicable remedy was seen as the construction of a public sewer system. Dilution of the ABS in the groundwater was then expected to follow.

From 1966 to 1970 Perlmutter and Koch[476] conducted a similar study on the distribution of methylene blue active substance (MBAS) and phosphate in the groundwater of Nassau County, Long Island, New York. The MBAS, a detergent constituent, entered the groundwater in the sewage effluent from several hundred thousand cesspools and septic tanks. The phosphate had a mixed origin. The MBAS was a significant problem in the public water supply aquifer, and the generally low concentrations of both constituents were not known to be toxic.

In a 1965 Federal Housing Administration publication, Wayman, et al.[477] concentrated on the mechanisms and physical principles of detergent (principally ABS) movement underground. The chemical characteristics of ABS, its movement with sewage effluent into groundwater, and its pollution tracing capabilities were analyzed.

Holloway[478] wrote a 1965 Texas Water Commission Bulletin on ABS in the groundwater of Rhineland, Knox County, Texas. The sources of the detergent contamination were found to be poor water well construction (drilled and dug wells without airtight covers) and domestic sewage from cesspools and septic tanks. Recommendations included the immediate plugging of polluting wells and the development of a public groundwater supply with the source remote from individual disposal facilities.

The California Department of Water Resources[479] published a 1965 report on the dispersion and persistence of synthetic detergents in the groundwater of San Bernardino and Riverside Counties. The report included chapters on the development of methods and sampling techniques, ABS concentrations in groundwaters, and the movement and degradation of ABS in groundwaters.

The fate and effect of nitrilotriacetic acid (NTA) both in groundwaters and in soil profiles overlying groundwaters were studied by Dunlap, et al.[480, 481] Sorption of NTA on soils slowed its movement into and through groundwaters, although sorption was not sufficient to prevent or greatly reduce potential pollution of groundwater by NTA used as a detergent builder. The infiltration characteristics of NTA in saturated (limited degradation) and unsaturated soils (rapid and complete degradation) were discussed, as were the possible metals produced by NTA which escaped such degradation. NTA degraded slowly in essentially anaerobic groundwater environments, resulting in production of $CO_2$, $CH_4$, and possibly other organic compounds.

A series of laboratory studies by Klein[482, 483] and Klein and Jenkins[484] evaluated the fate of various detergents in septic tank and oxidation pond systems. The biodegradability of polypropyl ABS and straight chain ABS, NTA, and carboxymethyloxysuccinate (CMOS) were examined, and emphasis placed on the relative removal rates of the detergents in septic tank-percolation fields versus oxidation ponds. In each study, the conditions necessary to avoid serious groundwater pollution problems were analyzed for possibility and practicality.

## NITRATES AND PHOSPHATES

Sources and methods of controlling nitrates and phosphates in groundwater supplies have been examined in detail.[485 to 493] Unusually high concentrations of nitrates in groundwater were linked to methemoglobinemia in infants and to animal health problems by Keeny[489] in 1970 and Winton, et al.[492] in 1971.

The sources of high nitrate concentration are many and varied. A 1970 University of Illinois sanitary engineering conference[485] cited inadequate biological sewage treatment systems, septic tanks, the harvesting of trees, ploughing up of root zones, and surface paving as factors hastening nitrate movement to groundwater. Sepp[486] in 1970 analyzed the nitrogen cycle in groundwater, and attributed high nitrate contents to agricultural practices and/or the land disposal of sewage (spreading or direct injection).

In 1970-71 Goldberg[488, 491] reviewed existing research, field, and laboratory studies of nitrogen sources, including atmospheric and geologic factors, rural and urban runoff (septic tanks), sewage, irrigation, animal feedlots, and industrial wastes. It was also reported that nitrate in a nonsalt form seemed to have a higher soil infiltration capacity than fertilizer salts of nitrogen. Keeny's[489] 1970 study concentrated on organic nitrogenous waste disposal, unwise fertilizer use, and percolation from feedlots as components of the nitrogen cycle. In 1971 Viets and Hageman[490] surveyed the general factors affecting nitrate accumulation in soil, water, and plants. The significance of geologic deposits, organic soil matter, agricultural wastes, domestic sewage, and commercial fertilizers was discussed.

Various mechanisms and procedures for nitrogen removal were described by Sepp,[486] Keeny,[489] and Lance[493] in 1970 and 1972. High nitrate groundwaters could be avoided by blending (dilution), proper aquifer selection, and proper well construction and sealing. Cropping, leaching, erosion, and volatilization can remove nitrogen from soils. Nitrogen removal from wastewater may be achieved by algae ponds, ion exchange processes (too expensive for large scale use), ammonia stripping, microbial denitrification, and electrodialysis. In promoting denitrification, direct injection of effluents was thought to be more effective than spreading.

Black[487] in 1970 presented an account of selected aspects of the behavior of soil and fertilizer phosphorus as they related to phosphate content of groundwaters. Chemical and geologic phosphorus cycles in the soil were traced, and the distributions of both inorganic and organic phosphorus in soils and groundwaters were discussed.

Numerous investigations of nitrates and phosphates in groundwaters of various states have been described. Kimmel[494] analyzed the nitrogen concentration in Kings County, Long Island, New York, groundwater from 1963 to 1971. Septic tank and cesspool effluents were thought to be the chief source of nitrates, but high nitrate levels were observed to continue even after widespread sewer construction in the area.

The nitrate contents of groundwater in New Castle and Kent Counties, Delaware, were studied by Miller[495] in 1972. The relationship among septic tank effluent disposal, geologic conditions, and water table depth was explored. Hazardous nitrate levels in one county required deep artesian wells until public sewers were constructed.

In 1956 Walker[496] reported on unusually high nitrate concentrations in wells tapping limestone aquifers in the Hopkinsville area of Kentucky. The contamination was attributed to human and animal wastes which moved long distances through aquifer crevices and openings. The necessity for safe well location far upslope from barnyards and houses and for effective sampling programs was stressed.

A 1971-72 study in Hartsville and Florence, South Carolina, by Peele and Gillingham[497] revealed excessive nitrate concentrations in groundwater and tile drainage effluent. Excessive nitrogen fertilizer application was the source; normal, safe fertilizer quantities for various crops were given.

The groundwater nitrate situation in Illinois has been detailed by Larson and Henley, [498] Dawes, et al.,[499] and Walker[500] since 1966. It appeared that about 25 percent of groundwater samples from shallow wells exceeded 45 mg/l $NO_3$. Primary causes of contamination were faulty well construction, domestic and industrial wastes, crop residues, decomposition of animal or plant tissue, and nitrogen fertilizer. No known practical and economical method of recovering excess nitrates existed, but membrane techniques and biological methods both had preventive potentials. Walker's[500] study focused on rural nitrate problems, recommended restricted quantities of fertilizer application, and suggested that waste disposal on farmlands be limited to the growing season of each year to exploit the storage capabilities of trees and plants.

Nitrate accumulation in Kansas feedlot and groundwaters was examined by Murphy and Gosch[501] for 1967-69. Regions of excessive nitrogen irrigation revealed large fluctuations in groundwater nitrate content. The study was inconclusive in relating the lack of nitrates in soil profiles to higher nitrate levels in underlying shallow aquifers; however, overall nitrate contents were higher in the winter months.

In 1967 Engberg[502] analyzed excessive nitrate concentrations in Holt County, Nebraska, well waters. Various localized contamination sources included barnyard and feedlot wastes, septic tanks, cesspools, silo seepage, excessive fertilization, and poor well construction. The threat of cyanosis to infants and health threats to animals were cited. Well site selection and construction criteria were considered, along with the possibility of deionizing units as nitrate control methods.

Wisconsin's nitrate situation was examined in three studies during 1968-70. Witzel, et al.[503] analyzed the nitrogen cycle in surface and subsurface waters, with emphasis on autotrophic and heterotrophic nitrification in various soil groups. Olsen[504] investigated the contribution of agriculture to nitrate increases, and stressed the effects of leaching and surface runoff. Control recommendations included limited nitrogen fertilizer use, crop covers during the growing season, crop rotation, and removal of unprotected manure during leaching periods. The possibility of anaerobic lagooning of manure as a safe and effective disposal method was also discussed. Crabtree[505] studied the nitrate contents of about 400 private, milk supplier, and nondiary farm wells around Marathon County, and found that about 40 percent exceeded safe nitrate levels of 45 mg/l.

In 1969 soil scientists reported in *Agricultural Research*[506] that no significant nitrate pollution of groundwater from fertilizer or feedlot operations was found in Northeastern Colorado. However, the studies indicated that excessive nitrate quantities could eventually reach the groundwater under heavily fertilized irrigated fields and feedlots. The type of land use along the South Platte River Valley did not appear to affect nitrate concentrations.

Taylor and Bigbee[507] in 1972 investigated fluctuations in nitrate concentrations to assess agricultural contamination in the semiarid regions of the Southwest. Regions studied included areas treated with nitrogenous fertilizers and subsequently irrigated and areas with varying animal densities compared to water usage.

The distribution of phosphorus in a fertilized and unfertilized Mexico soil was measured by Blanchar and Kao[508] in 1970-71. Phosphorus distributions, adsorption capacities, and solubility studies were conducted on various soil profiles.

Southern California nitrate problems were described by Navone, et al.[509] and the California Bureau of Sanitary Engineering[510] in 1963. Nitrates from fertilizer use, irrigation with reclaimed wastewater, and sewage disposal were observed in quantities exceeding 10 ppm $NO_3$-N in about 10 percent of 800 sampled wells.

In 1969 the Federal Water Quality Administration[511] presented a collection of eleven papers dealing with nitrate concentrations in subsurface agricultural wastewaters, sources of nitrates, and possible control or removal methods. The work concentrated on the San Joaquin Valley of California, but much of the information had general application. Included was a digital computer program developed by Shaffer, et al.[512] to model soil-water systems and to aid in planning management criteria for pollution control and nitrogen fertilizer programs.

Ward[513] in 1970 summarized nitrate groundwater investigations in six problem areas of California from 1953-68. Monitoring procedures were reviewed, and the validity of existing nitrate standards was considered.

In 1965 Stout, et al.[514] investigated high (100 ppm) nitrate concentrations in well waters of Grover City and Arroyo Grande, California. The nitrate probably originated from decomposition of native plant covers, but agricultural fertilizers, septic tanks, and sewers also contributed to the situation. Well construction criteria and restricted pumping techniques were emphasized as protective measures.

The California Department of Water Resources[515] and the staff of *Environment*[516] in 1968-69 reported on serious groundwater nitrate in the state, particularly at Delano and McFarland. Methemoglobinemia in infants was threatened, due to nitrates from sewage discharge, fertilizers, and irrigation. As a temporary solution, Delano pumped only wells with low nitrate levels, but long term control required proper drainage of irrigated land along with distillation and treatment of agricultural runoff.

The Fresno-Clovis metropolitan area of California has been investigated by Nightingale[517] and Schmidt[518, 519] in 1970-72. From 1950 to 1967 the salinity and nitrate contents of well water in urban and agricultural areas had increased, constantly in the urban zone and irregularly in the agricultural zone. Chemical hydrographs were employed to discover the distribution and potential sources of nitrates in groundwater. Highest nitrate contents were found in the shallower parts of the aquifer. Primary sources of nitrates included septic tanks, sewage treatment plants, percolation ponds, winery wastewater ponds, and agricultural fertilizers.

Nightingale[520] in 1972 also presented a study of nitrates in the root zone and in groundwaters beneath irrigated and fertilized crops in the Fresno County area. Some localized spots of high groundwater nitrates were evident, mostly associated with sewage disposal systems in urban areas; in general, fertilizer practices appeared safe.

In 1970 and 1972 Willardson, et al.[521, 522] reported on a field study of the effectiveness of submerged drains in reducing nitrates near Firebaugh in the San Joaquin Valley of California. Nitrates in the area's groundwater were attributed to nitrogen fertilizer use in irrigation. Denitrification occurred in saturated soil where there was ample organic carbon available for bacterial metabolism. Results indicated a high degree of denitrification and dilution of high nitrate groundwater in the test area.

In 1972 Pratt, et al.[523, 524] described a field study of nitrates in deep soil profiles and their relation to fertilizer rates and leaching volume. The study area was situated in the Santa Ana River Basin of California. Saturation extracts and soil solutions were measured and transit times for water to move 30 meters in the unsaturated zone to the water table beneath commercial citrus groves were calculated.

In 1973 Ayers and Bronson[525] documented a field study of groundwater nitrates in the Upper Santa Ana River Basin. This detailed report included tables of groundwater flow rates, and considered nitrate reduction by control of fertilizer use and application, water use, and waste disposal.

## HEALTH

There have been a number of studies regarding the public health aspects of groundwater pollution. Vogt[526] reported on a 1959 infcctious hepatitis epidemic in Posen, Michigan, caused by septic tank drainage through limestone to shallow wells. The virus transmittal resulted in 89 reported cases of infectious hepatitis, and construction of a municipal water supply was seen as the best means of control. Bacterial, chemical, and radiological contamination of the aquifer drinking water in Monroe County, Michigan, was the subject of a 1960 study by Hancock[527]. Twelve disinfectants capable of killing disease producing organisms in water were listed, and two consecutive safe coliform tests were recommended before an aquifer is judged safe for drinking water. The roles of the government and the water well industry in the control of well location, construction, and operation were also stressed.

Two outbreaks of waterborne intestinal disease in California prior to 1966 led Foster and Young[528] to study means of replenishing underground basins with good quality water. Subsurface filtration alone may not effectively remove pathogens and toxic chemicals; the efficacy and desirability of chlorination were discussed. In addition, waterborne outbreaks of disease since 1920 were reviewed; over 50 percent were found to be due to either contamination of a well supply or cross-connections and other hazards in the distribution system.

A 1969 report by Robeck[529] also discussed disease outbreaks caused by contaminated groundwater and the problems of controlling and monitoring groundwater quality. Microbial contamination can be caused by recharge of large basins with reclaimed sewage, and coliforms are not always useful indicators of such contamination. Salmonella outbreaks have occurred without detectable coliforms, and many media capable of passing virus particles will also filter out coliforms. The difficulties posed by nitrates in waste water and the resulting public health dangers were also discussed. The author concluded that chlorination of all groundwater for domestic use is the best assurance of microbial control.

An apparent case of pesticide poisoning in 1969 in Central Idaho was the subject of a field report by Benson.[530]   Water samples indicated E. coli at a concentration of 17.2/cc and the insecticide Thimet in a shallow dug well used for water supply. The apparent source of the insecticide was a drainage ditch within 30 feet of the well.   A new deeper well was drilled, which yielded water of satisfactory quality.

## MISCELLANEOUS

Hodges, et al.[531] conducted a field study in 1961-62 of gas and brackish water in the fresh water aquifers of the Lake Charles area of Southwest Louisiana.   Methane concentrations up to 82 ppm (0.2 ppm is normal) were found in the groundwater, partly a result of the generation of "marsh gas" within the aquifer itself, and perhaps partly due to the movement of petroliferous gas through fault zones, abandoned well holes, and well blowouts. The abandonment of some industrial wells in the mid-1950s resulted in upward saltwater movement and helped explain the presence of brackish groundwater of generally low chloride content.

A 1962-64 study by Harder, et al.[532] of the methane problem in a wider area of Southwest Louisiana revealed methane concentrations of 0-127 ppm.   Three hundred sampling wells, screened at various depths, showed higher methane concentrations in the southern area of the study near three oil and gas producing fields.   Most of the methane in the groundwater was generated by organic debris within the aquifers, but some probably originated in the oil and gas sands beneath the aquifers and moved upward through defective well casings in existing or abandoned wells.   The theoretical hazards to groundwater users from explosive methane-air mixtures was also discussed.   To control the problem the authors recommended strict well construction and abandonment regulations as well as thorough aeration of the groundwater before use.

A 1969 report of the U.S. Federal Water Pollution Control Administration[533] dealt with the problem of sewer line leaks and water infiltration.   New, more effective sealants were developed, and the costs and effectiveness of the various equipments and materials investigated were presented.   New equipment designs were also described and recommended.

Grossman[534] in 1970 reported on waterborne styrene in a crystalline bedrock aquifer in the Gales Ferry area, Ledyard,

Connecticut. Following shallow burial in 1960, the styrene moved downward into the aquifer. Movement from two unspecified sources of contamination to cones of depression at six domestic wells was along foliation joints and cross joints and exceeded 300 feet. Removal of the sources of contamination produced water of satisfactory chemical quality within two years.

In 1970 Dixon and Hendricks[535] presented a water quality simulation model in conjunction with a hydrologic simulation model to aid in development and planning aspects of aquifer resource management. The model represented water quality changes in both time and space in response to changing atmospheric and hydrologic conditions and to time-varying waste discharges at various points in the system. Procedural guidelines were also given to assist in the development of water quality simulation models as tools for use in the quality-quantity management of a hydrologic unit.

Folkman and Wachs[536] conducted a laboratory study of the processes occurring when effluents of stabilization ponds are used in artificial groundwater recharge. Algae concentration experiments studied the filtration of chlorella through columns of dune sand. The report discussed the changes in relative concentration of algae as a function of depth, and demonstrated the increases in filtration efficiency due to increased cation concentration in the water and lower water velocities.

In 1972 Aley, et al.[537] surveyed the problem of groundwater contamination and sinkhole collapse in Missouri. Numerous case histories of land collapses of lagoons and impoundments were presented. Spores for tracing groundwater in limestone were discussed, and thorough hydrogeological investigations of soluble rock and limestone terrain were recommended as means of control.

# CHAPTER X

# EVALUATING POLLUTION

## GEOLOGY AND TRACERS

The geologic and hydrologic factors involved in the ground disposal of wastes have been analyzed by Ferris[538] in 1951 and by Theis[539] in 1955. Ferris concentrated on the use of injection wells for industrial waste disposal and emphasized proper well drilling, casing, construction, and plugging, as well as hydrologic isolation of the disposal formation from aquifers to prevent leakage. Theis focused on favorable and unfavorable features of groundwater circulation bearing on the problem of underground waste disposal (especially with regard to radioactive wastes). Both studies concluded that detailed knowledge of local and regional geology and hydrology were prerequisites to safe underground waste storage.

In 1964 LeGrand[540] presented a system for evaluating the contamination potential of waste disposal sites. The method was based on three categories of site geology and on weighted values of the water table, sorption, permeability, water table gradient, and distance to point of use. Examples were cited and alternative proposals offered.

Numerous studies have also been done on the general problem of geologic controls over groundwater.[541 to 544] Since contaminated groundwater is subject to the same physical laws as pure water, geology actually controls groundwater contamination. The extent to which a contaminant affects groundwater depends to a large extent on the geologic factors affecting groundwater movement and the capacity of rock materials to absorb and adsorb the contaminant. The studies have thus detailed these geologic principles involved in contaminant entry, removal, and dispersion in groundwater, some illustrated by case histories.[542, 544]

In 1964 Stewart, et al.[545] described the results of a geologic and hydrologic investigation at the site of the Georgia Nuclear Laboratory in Dawson County, Georgia. The purpose of the study was to determine: (1) the occurrence, rate, and direction of movement, discharge, and recharge of groundwater; (2) water quality and quantity; and (3) the effects of liquid waste disposal on the groundwater near the radiation sites.

Maxey and Farvolden[546] presented a 1965 discussion on the specific hydrogeologic factors in problems of groundwater contamination in arid lands. They concluded that the suitability of hydrogeologic units for any water supply or waste disposal operations depended primarily on the position within the hydrologic system and secondarily on physical properties. The ideal hydrologic system in arid lands involved a recharge area in mountains and a discharge area in lowlands. The compatability of waste disposal methods to the groundwater flow system at the Nevada Test Site and at Las Vegas were compared and contrasted.

In 1967 Morris[547] described the use of chemical and radioactive tracers in studies of the geology and hydrology of the basalt terrane at the National Reactor Testing Station in Idaho. Rates of groundwater flow were determined by tracers of salt, sodium fluorescein dye, and tritium.

In 1967 Marine[548] reported on the use of a tracer test to verify an estimate of the groundwater velocity in fractured crystalline rock at the Savannah River Plant near Aiken, South Carolina. The storage of high-level radioactive wastes in unlined tunnels in the crystalline rock had been found to be technically feasible, and hydraulic estimates of groundwater velocity had been made. Using tritium as a tracer, the measured groundwater velocity was discovered to be 2.5 times the predicted average velocity.

In 1970 Webster, et al.[549] described a further two-well tracer test in fractured crystalline rock under the Savannah River Plant near Aiken, South Carolina. A pulse injection of tritium was made to flow from an injection well 1,765 feet to a discharge well, and the duration of the test was two years. The concentration of tritium arriving at the discharge well agreed with predicted calculations based on fluid dispersion in a homogeneous medium.

In 1971 Armstrong, et al.[550] reported on the use of tritiated water as a tracer to follow the path of leach liquids as they flowed through a copper mine dump. Semipermeable layers within the dump restricted the vertical movement of water, and optimization of the leaching process depended on knowledge of this restriction. Tritiated water was injected into the dump, and the leach liquid was sampled at natural surface outflows and through a series of wells. More than 3,300 samples were analyzed by liquid scintillation counting or by gas counting. The data obtained permitted calculation of flow paths, recycling times, total fluid volumes, and estimates of retention times in various portions of the dump.

## POLLUTION TRAVEL

Many studies have been conducted relating to the general hydrologic and geologic factors involved in the movement of contaminants through soil and within the aquifer. In 1964 Brown[551] discussed the hydrologic factors affecting the travel of fluid waste from a disposal site to the water table and within the aquifer. The fluid movement was shown to depend on the location and extent of all pervious and impervious materials in the zone. In addition, the choice of disposal technique (disposal pit versus recharge well) was crucial in predicting the time required for the fluid waste to reach the water table.

In 1967 McGauhey and Krone[552] presented a detailed survey on the possibility of engineered wastewater systems exploiting the soil mantle of the earth as a means of wastewater treatment. Successful design and operation of such a system depended on a soil's infiltrative capacity approaching its percolative capacity. The ability of a soil to remove or transmit bacteria, viruses, and chemicals was reviewed. Finally, ten areas of needed research were outlined.

The effects of temperature and density gradients on the movement of contaminants in groundwater were studied by Henry[553] in 1968. A horizontal lateral temperature gradient was imposed on a tube of square cross-section containing water-saturated sand. A convection current increased in strength as the temperature gradient increased. The effective value of thermal diffusivity was about fifty times larger than the expected value because of dispersion caused by water movement through interconnected interstices of the sand.

In 1969 Boyd, et al.[554] conducted investigations of the basic mechanisms by which surface pollution may gain entrance to groundwater. The collected data showed that while moisture and nutritive values of various Colorado soil types were important for bacterial survival, microbial overpopulation was a major cause of bacterial death. Additional data revealed that the size of sand granules and the specific type of ion present in bacterial suspensions greatly affected the mobility of bacteria through sand columns.

A 1969 report by Champlin[555] centered on the fundamental mechanisms by which trace metals and organic compounds were fixed or immobilized by soil or earth sediments from water moving through aquifers.  Dilute water suspensions were passed through packed sand beds, and retention by the sand bed of radioactivity added to water, suspended bacteria, and suspended clay was examined.  The research established that significant movement of ionic matter through porous beds of soils could take place in the form of dilute suspensions at low salt concentrations, but that direct transfer to free ions through packed beds at low dissolved salt concentrations was unlikely.

In 1971 Champlin[556] described further experimental data which showed a close relation between the relative movement of both trace ions and particles and the overall concentration of common salts dissolved in groundwater.  Most importantly, the spatial stability of fine particles such as the sesquioxides and the clays in formations was dependent on the nature of the anionic portion of the dominant salt in solution.  It was thought that fine particles tagged with almost any radioactive or activatable ion might become an invaluable tool in tracing subsurface movements of fluids.

Hajek[557] presented in 1969 a study on predicting the performance of soil as a wastewater disposal and water reclamation resource.  Significant wastewater parameters included:  pH, pollutant form and concentration, concentration of accompanying ions, temperature, and volume disposal characteristics.  Significant soil parameters were:  bulk density, grain size distribution, mineral composition, exchange capacity, and resident exchangeable cations. Batch equilibrium and dynamic soil column studies generally characterized the interrelationships between soil and wastewater. These data could be employed to predict pollutant migration rates and concentration distributions.

In 1970 Romero[558] reported on guidelines for safe distances between domestic and/or food processing wells and potential or existing sources of groundwater pollution.  Safe and effective use of the filtering capacity of the soil mantle as a wastewater treatment system was stressed.  The report concluded that no one set of "safe distances" was adequate and reasonable for all locales and conditions.

Many more studies have been conducted since 1950 on underground movement of specific pollutants, often in specific locations. Butler, et al.[559] in 1954 summarized studies of bacterial and chemical pollutant travel in the arid Southwest for both above and below the water table conditions. Above the water table, bacteria travelled only limited distances in both fine and coarse soils, while chemical pollutants were little altered by passing through as much as thirteen feet of unsaturated soil. Below the water table, coliform organisms in groundwater travelled up to 232 feet, while chemical pollutants were found to travel farther (up to several miles) and faster.

Using various test organisms, Fournelle, et al.[560] in 1957 reported on a study of the lateral movement of simulated bacterial and chemical pollutants in shallow groundwater in Anchorage, Alaska. The field study showed that dye uranin and streptococcus zymogenes were very effective in determining the direction of groundwater flow and the extent of pollutant travel through the groundwater, even after three years. The authors also developed criteria for the selection of effective test organisms for such studies.

Sampayo and Wilke[561] detailed a 1961-62 study of the effects of recharged water on groundwater temperatures, phosphate concentration, and the direction of groundwater flow in West Lafayette, Indiana. The influence of storm and warm air conditioning waters was found to be localized to a small area (1,000 feet) surrounding the recharge pit.

In 1963 Page, et al.[562] summarized soil column studies on the comparative effectiveness of coarse grained, fine grained, and colloid coated soils in removing ABS, dissolved solids, and bacteria from sewage effluent under saturated flow conditions. The coarse sand and sandy loam removed about 90 percent of the bacteria from sewage within a few feet (although additional travel did not always remove remaining bacteria), but the dissolved solids and ABS were virtually unaffected by filtration. The use of colloidal alumina to remove ABS or bacteria was considered economically infeasible, and the coarser sand was preferred as a pollution filter to remove bacteria to prevent clogging.

In a similar laboratory study in 1967, Young et al.[563] analyzed the ability of four Oahu clay-type soils to remove ammonia, ABS, and coliforms from water percolating continuously through saturated soil columns. The results were not conclusive because of small soil samples and loading procedures, but preliminary

tests on a 30-inch column of Wahiawa soil showed an initial coliform reduction of about 90 percent.

In 1967 Kumagi[564] reported on a laboratory study of infiltration and percolation of sulfides and sewage carbonaceous
matter.  Utilization of simulated cesspool lysimeters resulted in
free percolation of certain odorous compounds and excellent COD
removals under presumably anaerobic conditions, contrary to findings in similar studies.  Soil columns were found more effective
than sand columns for sulfide removal, but under acid conditions
sulfide breakthrough was evident in both soil and sand columns.
All columns exhibited the characteristic nonlinear relationship between filtration and percolation rates and the hydraulic gradient.

Ishizaki, et al.[565] studied the passage of an organic-rich
liquid through cracks in Hawaiian basaltic lavas.  Their 1967 report determined the permeability and porosity values for various
portions of a basalt.  The flow of organic-rich liquids through
such cracks, similar to nonbiodegradable liquids, exhibited a decrease in flow initially and continued the trend for as long as
220 hours.  The clogging phenomenon was dependent upon microbial activity and food supply in the sewage.

In 1968 Krone[566] presented a study of the movement of
pathogenic organisms through soils recharged with contaminated
water.  The physical and biological characteristics of pathogens
were discussed, along with various processes of soil filtration.  The
straining of pathogens at the soil surface and the sorption of
viruses near the surface were the most effective limitations on
pathogen travel.  Therefore, a soil containing clay was recommended for irrigation with treated sewage.  Secondary treatment
and chlorination of the sewage were also recommended for aes-
thetic reasons.

A test drilling program was conducted by Crosby, et al.[567]
in a drain field area of the Spokane Valley, Washington, to study
the movement of pollutants in glacial outwash deposits subjected
to extreme pollutant loads.  Very dry soils were found about 30
feet below the drain field, and it was concluded that most of
the waters were being dispersed laterally by capillary mechanisms.
Chemical pollutants were found to travel with moisture fronts,
but fine materials were determined to be very effective in filtering bacteria within a few feet of the leach bed.

Further study at the same site by Crosby, et al.[568] resulted
in a 1971 data analysis and literature review of the very high
prevalent moisture tensions in the drain field environment.  The

high moisture stress was seen as contributing to the rapid removal
of bacteria during filtration.   Nitrates and chlorides in the groundwater seemed to be unrelated to percolating wastewaters, and no
bacterial pollutant threat to area well waters from the drain field
operations was discovered.

Scalf, et al.[569,570] presented data on the movement, adsorption, and release of nitrate and DDT under actual well recharge conditions in the Ogallala aquifer at Bushland, Texas, in
1968.   Tritiated water used as a tracer revealed that 94 percent
of the recharge water was recovered in twelve days of withdrawal
by pumping.   Chemical and hydrologic analyses also disclosed
that the movement of nitrate was not the same as the tritiated
water, and that the DDT apparently was adsorbed by the aquifer
material close to the recharge well since very little was recovered
during pumping.

In 1972 Tilstra, et al.[571] described studies on the removal of
phosphorus and nitrogen from wastewater effluents by induced
soil percolation.   Induced percolation achieved good stabilization
for phosphorus, but the results for nitrogen were poor.   Superior
results were achieved when aerobic conditions existed in the soil.

Wentink and Etzel[572] in 1972 analyzed soil column tests
dealing with the removal of metal ions by soils.   The ion exchange capacities of three soils were observed to effectively remove
copper, chrome, zinc, nickel, and cadmium ions.   The exchange
capacity of a soil increased as its clay mineral content increased
and as its particle size decreased.

In 1973 Allen and Morrison[573] reported on a field investigation of percolating leachfield effluents in the mountainous
Parvin Lake area of Colorado.   The direction and rate of movement of bacteria-laden effluent were mainly affected by the anisotropy of the bedrock fracture patterns, and percolating effluent
was observed to have travelled distances of over several hundred
feet through fractured bedrock.   Consequently, even moderate
percolation rates and large distances between water wells and conventional waste disposal units may not guarantee potable groundwater in mountainous areas of crystalline bedrock.

## MONITORING

In 1956 the California Department of Public Works[574] reported on a general program of groundwater quality monitoring

in California.  Groundwater quality data were summarized, along with an outline of the state monitoring philosophy.

Bookman and Edmonston[575] summarized the water-quality-monitoring and -research programs of six city, county, and state agencies in the San Gabriel River System, Los Angeles County, California, in 1962.  Agency permit requirements were found to be satisfactory methods of waste disposal control, but reclaimed sewage, along with well construction and sealing, techniques remained a problem.

Pomeroy and Orlob[576] presented a 1967 discussion of water quality standards in California.  Special monitoring problems of groundwater pollution included spatial and temporal quality variations, groundwater quality constituents (dissolved minerals, bacteria, radioactivity, temperature, oxygen content), and minimal surveillance requirements (observation wells versus outflow sampling).  In addition, the report contained a checklist of water quality indicators for various uses of water.

A 1969 report by Moreland and Singer[577] suggested selective analyses for obtaining specific types of groundwater quality data in the Orange County Water District, California.  In practice, water samples were collected periodically from 272 wells in the area.  Chloride and electrical conductivity measurements were made on samples from aquifers susceptible to sea water intrusion.  Sulfate, bicarbonate, and nitrate determinations were made on samples from aquifers underlying the forebay area.  Finally, sodium, sulfate, chloride, and boron determinations and electrical conductivity measurements were taken on samples from aquifers used as a source for irrigation water.

Water Resources Engineers, Inc.,[578] conducted an investigation of the salt balance in the Upper Santa Ana River Basin in Southern California.  A two-volume report in 1970 detailed the sources and control methods of the "controllable" salt accretion of the area.  Effective hydrologic and water quality models for the region were developed, including a complete data management system.  In addition, heavy emphasis was placed on the location of monitoring wells and the scheduling of observations in the basin.

The most recent evaluation of water quality monitoring programs in California was done by the State Water Resources Control Board[579] in 1971.  State and federal agency information gathering programs, methods, objectives, and problems were discussed. The study stressed the need for a statewide integrated agency monitoring program geared toward a centralized data and water resources management system. Increased use of automatic recording and aircraft sensors was recommended. The need for better equipped laboratories and more efficient self-monitoring programs was also expressed.

Methods employed by the U.S.G.S. to collect, preserve, and analyze water samples were reviewed by Rainwater and Thatcher [580] in 1960. The selection of sampling sites, frequency of sampling, field equipment, preservatives and fixatives, analytical techniques of water analysis, and instruments were among the topics discussed. Seventy-seven laboratory and field procedures were listed for determining 53 various water properties.

A 1972 Federal Interagency Work Group[581] study focused on recommended methods for acquisition of groundwater pollution data. Detailed summaries were provided of acquisition methods for various parameters and sampling systems with respect to biologic, bacteriologic, chemical, and physical water quality data, and for automatic quality monitoring systems.

In 1973 the Environmental Instrumentation Group of the University of California[582] presented a detailed report on the physical and operating characteristics and specifications of presently available water quality monitoring instruments. Various instrumentation methods were critically compared, and recommendations were made on promising methodology and the development of new instrumentation. The report contained numerous references, and contemplated future expanded discussion of various water quality parameters in addition to metallic, biological, oil and grease, and physical parameters.

More limited studies in this field have included reports on computer methods, management systems, and water quality models, along with estimation theory analyses of water quality monitoring problems. Morgan, et al.[583] in 1966 explained the advantages of a digital computer in storing, retrieving, and manipulating water quality data. Included were examples of tables, Stiff and Piper diagrams, and maps produced by the computer.

In 1971 Simpson, et al.[584] presented a paper which sought to define problems in space-time sampling of pollutants in aquifers and to give guidelines detailing the circumstances under which mean value or deterministic sampling models were justified. Two general classes of problems were distinguished: input identification and system identification. In addition, a finite-state machine model was proposed for space-time sampling of aquifer pollutants, and some theoretical sampling scheme problems were discussed.

A 1971 study of the Hanford groundwater basin in Washington by Cearlock[585] led to the development of a systems approach to the management of area radionuclide pollution problems. A man-machine interactive computer system was employed to produce hydraulic models of the groundwater flow in saturated and unsaturated soils and water quality models of waste movement through subsurface soils.

The analysis, modeling, and forecasting of stochastic water quality systems were researched by Lee[586] in 1972. Time series analysis, optimal and non-linear filtering, and estimation theory approaches to the problem were discussed. The study contained numerous references and examples of mathematical representations of various pollution parameters.

Zaporozec[587] in 1972 surveyed methods and techniques of graphical and numerical interpretation of water quality data. Classification, correlation, analytic, synthetic, and illustrative methods were summarized and charted.

In 1973 Moore[588] analyzed the application of estimation theory to the design of water quality monitoring systems. Filtering techniques provided a potentially valuable methodology to this end. For example, the "best" sampling program could be selected from a group of feasible measirement systems by sequentially minimizing a cost function subject to constraints on the uncertainty of estimates. Trade-offs might then be necessary between spatial and temporal frequencies of sampling.

Still further groundwater quality monitoring studies have centered on specific sampling equipment or measurement techniques. McMillion and Keeley[589] in 1968 described the design and specifications of new portable pumping equipment which could sample to depths of 300 feet at pumping rates of 7-14 gallons/minute. The equipment could investigate chloride reduction rates in an aquifer under the influence of selective pumping techniques and could trace pollutants in a fresh water aquifer under alternate recharge and pumping conditions.

In 1968 LeGrand[590] discussed general groundwater quality problems related to test and monitor wells. In order to achieve effective monitoring with optimum results, the author concluded that improved technology in determining the distribution of contaminated groundwater and synthetic hydrogeologic frameworks with adequate data was essential.

In 1969 Warner[591] reported on attempts to detect and outline zones of groundwater contamination by earth resistivity measurements where a resistivity contrast exists between contaminated and uncontaminated groundwater. Three surveys over Long Island, New York sites (septic tanks were a source of groundwater pollution) were particularly successful, as was one Western Texas survey (groundwater pollution due to oil field brine disposal pits). The earth resistivity measurement method was found to be useful for rapid economical surveys of large land areas, and for monitoring water level and water quality changes in large topographically uniform areas where unconfined aquifers exist.

In 1970 Peterson and Lao[592] reported on the use of spontaneous potential, resistivity, and electrical conductivity well logs to measure groundwater flow zones, location, depth, and the chemical constituents of the groundwater in Hawaii. Data from the fluid conductivity logs combined with temperature data also aided in interpretation of the Ghyben-Herzberg lens relationship.

Turcan and Winslow[593] presented a 1970 study of techniques for evaluating electrical logs to aid in the estimation of the volume and distribution of saline groundwater in Louisiana. In addition, well yields and altitudes of groundwater salinity interfaces were estimated by quantitative analysis of borehole geophysical logs.

In a similar investigation, Brown[594] reported in 1970 on techniques for calculating groundwater quality from calibrated geophysical logs in a Norfolk, Virginia, test well 2,500 feet drop. Methods for approximating the dissolved solids and chloride content of the groundwater were specifically detailed.

In 1972 Foster and Goolsby[595] summarized the results of a field investigation at Pensacola, Florida. Two monitor wells were constructed to discover the continuing effects of the injection of liquid chemical waste by Monsanto into the lower Floridan aquifer. Chemical analyses of water samples revealed highly saline groundwater, increasing in salinity with depth. The report included a detailed description of the construction of the monitor wells.

# REFERENCES

References 1 to 595 are cited in the text, while references 596 to 661 are not.

1. University of California Sanitary Engineering Research Lab., *Studies in Water Reclamation,* Univ. of California Sanitary Engineering Research Lab. Technical Bulletin 1, 65 pp, 1955.

2. Popkin, R. A., and T. W. Bendixen, "Improved Subsurface Disposal," *Jour. Water Pollution Control Fed.,* Vol. 40, No. 8, pp 1499-1514, 1968.

3. Tchobanoglous, G., and R. Eliassen, "The Indirect Cycle of Water Reuse," *Water and Wastes Engineering,* Vol. 6, No. 2, pp 35-41, 1969.

4. Bouwer, H., "Water-Quality Improvement by Ground Water Recharge," *Agricultural Research Service,* Report 41-147, pp 23-27, 1969.

5. Born, S. M., and C. A. Stephenson, "Hydrogeologic Considerations in Liquid Waste Disposal," *Jour. Soil and Water Conservation,* Vol. 24, No. 2, pp 52-55, 1969.

6. Martin, W. P., "Controlling Nutrients and Organic Toxicants in Runoff, " *Water Pollution by Nutrients — Sources, Effects and Control,* Water Resources Research Center, Univ. of Minnesota, Minneapolis, WRRC Bulletin 13, pp 39-47, June 1969.

7. Dvoracek, M. J., and R. Z. Wheaton, "Does Artificial Ground Water Recharge Contaminate Our Ground Water?" *Relationship of Agriculture to Soil and Water Pollution,* Cornell Univ. Conf. on Agricultural Waste Management, pp 69-76, 1970. (NTIS: PB-195 304)

8. Pennsylvania Bureau of Water Quality Management, *Spray Irrigation Manual,* Publication No. 31, Pennsylvania Dept. of Environmental Resources, Harrisburg, 49 pp, 1972.

9. Bernhart, A. P., "Protection of Water-Supply Wells from Contamination by Wastewater," *Ground Water,* Vol. 11, No. 3, pp 9-15, 1973.

10. Pennypacker, S. P., et al, "Renovation of Wastewater Effluent by Irrigation of Forest Land," *Jour. Water Pollution Control Fed.,* Vol. 39, No. 2, pp 285-296, 1967.

11. Sopper, W. W., "Waste Water Renovation for Reuse: Key to Optimum Use of Water Resources," *Water Research,* Vol. 2, No. 7, pp 471-480, 1968.

12. Anon., "Pollution-Free Sewage Disposal," *Ground Water Age,* Vol. 7, No. 10, pp 21-22, 27-28, 1973.

13. Bendixen, T. W., et al, "Ridge and Furrow Liquid Waste Disposal in a Northern Latitude," *Jour. Sanitary Engineering Div.,* Amer. Soc. of Civil Engineers, Vol. 94, No. SA 1, pp 147-157, 1968.

14. Ketelle, M. J., "Hydrogeologic Considerations in Liquid Waste Disposal, with a Case Study in Southeastern Wisconsin," *Technical Record,* Vol. 3, No. 3, Southeastern Wisconsin Regional Planning Commission, 39 pp, 1971.
    (NTIS: PB-205 951)

15. Chaiken, E. I., et al, "Muskegon Sprays Sewage Effluents on Land," *Civil Engineering,* Vol. 43, No. 5, pp 49-53, 1973.

16. Harvey, E. J., and J. Skelton, "Hydrologic Study of a Waste-Disposal Problem in a Karst Area at Springfield, Missouri," *Geological Survey Research 1968,* U.S. Geological Survey Prof. Paper 600-C, pp C217-C220, 1968.

17. Brown, R. F., and D. C. Signor, "Groundwater Recharge," *Water Resources Bulletin,* Vol. 8, No. 1, pp 132-149, 1972.

18. Anon., "'Overland-Flow' Irrigation System Solves Campbell Soup's Texas Disposal Problem," *Water and Sewage Works,* Vol. 120, No. 4, pp 84-86, 1973.

19. Bouwer, H., "Returning Wastes to the Land, A New Role for Agriculture," *Jour. Soil and Water Conservation,* Vol. 23, No. 5, pp 164-169, 1968.

20. Bouwer, H., "Putting Waste Water to Beneficial Use — The Flushing Meadows Project," *Proceedings of the 12th Annual Arizona Watershed Symposium*, Phoenix, Arizona, pp 25-30, Sept. 1968.

21. Bouwer, H., "Groundwater Recharge Design for Renovating Waste Water," *Jour. Sanitary Engineering Div.*, Amer. Soc. of Civil Engineers, Vol. 96, No. SA 1, Paper 7096, pp 59-74, 1970.

22. Bouwer, H., "Water Quality Aspects of Intermittent Systems Using Secondary Sewage Effluent," *Artificial Groundwater Recharge Conference Proceedings*, Vol. 1, The Water Research Assoc., Medmenham, England, pp 199-217, June 1971.

23. Bouwer, H., et al, "Renovating Sewage Effluent by Ground-Water Recharge," *Hydrology and Water Resources in Arizona and the Southwest*, Proceedings of Arizona Section-Amer. Water Resources Assoc. and the Hydrology Section-Arizona Academy of Science, Tempe, Vol. 1, pp 225-244, 1971.

24. Bouwer, H., et al, *Renovating Secondary Sewage by Groundwater Recharge with Infiltration Basins*, U.S. Environmental Protection Agency, Water Pollution Control Research Series 16070-DRV, 102 pp, March 1972.
    (NTIS: PB-211 164)

25. Wilson, L. G., et al, *Dilution of an Industrial Waste Effluent with River Water in the Vadose Region During Pit Recharge*, Meeting Paper No. 68-727, Amer. Soc. of Agricultural Engineers, 1968 Winter Meeting, Chicago, 26 pp, 1968.

26. Stone, R., and W. F. Garber, "Sewage Reclamation by Spreading Basin Infiltration," *Proceedings Amer. Soc. of Civil Engineers*, Vol. 77, No. 87, 20 pp, 1951.

27. California State Water Pollution Control Board, *Field Investigation of Waste Water Reclamation in Relation to Ground Water Pollution*, Publication No. 6, 124 pp, 1953.

28. Stone, R., "Land Disposal of Sewage and Industrial Wastes," *Sewage and Industrial Wastes*, Vol. 25, No. 4, pp 406-418, 1953.

29. California Dept. of Water Resources, *Reclamation of Water from Sewage and Industrial Wastes — Progress Report, July, 1953-June 30, 1955*, Bulletin 68, 24 pp, 1958.

30. McMichael, F. C., and J. E. McKee, *Wastewater Reclamation at Whittier Narrows,* California State Quality Control Board, Publication No. 33, 100 pp, 1966.

31. Doneen, L. D., "Influence of Native Salts in the San Joaquin Valley on the Quality of the Ground Water," *Proceedings of Symposium on Agricultural Waste Waters,* Report 10, Univ. of California Water Resources Center, pp 61-65, 1966.

32. Matthews, R. A., and A. L. Franks, "Cinder Cone Sewage Disposal at North Lake Tahoe, California," *Water and Sewage Works,* Vol. 118, No. 1, pp 2-5, 1971.

33. Boen, D. F., et al, *Study of Reutilization of Wastewater Recycled Through Groundwater,* 2 Vols., U. S. Environmental Protection Agency Report EPA-16060-DDZ-07/71, 330 pp, July 1971. (NTIS: PB-190 790; PB-190 791)

34. Boen, D. F., et al, *Reutilization of Wastewater Recycled Through Groundwater,* Vol. 1, Progress Report to Fed. Water Pollution Control Admin. for Project 16060, 25 pp, Dec. 1968. (NTIS: PB-190 790)

35. Boen, D. F., et al, *Reutilization of Wastewater Recycled Through Groundwater,* Vol. 2, Progress Report to Fed. Water Pollution Control Admin. for Project 16060, 87 pp, Dec. 1969. (NTIS: PB-190 791)

36. Young, R. H. F., et al, "Wastewater Reclamation by Irrigation," *Jour. Water Pollution Control Fed.,* Vol. 44, pp 1808-1814, 1972.

37. California State Water Pollution Control Board, *Report on the Investigation of Leaching of a Sanitary Landfill,* Publication No. 10, 92 pp, 1954.

38. Merz, R. C., "The Effects of the Sanitary Landfill on the Ground Water at Riverside," *Western City,* Vol. 31, No. 4, pp 46-48, 1955.

39. California State Water Pollution Control Board, *Effects of Refuse Dumps on Ground Water Quality,* Publication 24, 107 pp, 1961.

40. California State Water Quality Control Board, *In-Situ Investigation of Movements of Gases Produced from Decomposing Refuse,* Publication 31, 211 pp, 1965.

41. California Department of Water Resources, *Ground Water Basin Protection Projects: Sanitary Landfill Studies, Appendix A: Summary of Selected Previous Investigations,* Bulletin 147-5, App. A, 115 pp, 1969.

42. Coe, J. J., "Effect of Solid Waste Disposal on Groundwater Quality," *Jour. Amer. Water Works Assoc.,* Vol. 62, No. 12, pp 776-783, Dec. 1970.

43. Hughes, G. M., *Selection of Refuse Disposal Sites in Northeastern Illinois,* Illinois State Geological Survey Environmental Geology Note No. 17, 18 pp, Sept. 1967.

44. Hughes, G. M., et al, *Hydrogeology of Solid Waste Disposal Sites in Northeastern Illinois,* U.S. Dept. of Health, Education, and Welfare, Bureau of Solid Waste Management, Cincinnati, 137 pp, 1969.

45. Landon, R. A., "Application of Hydrogeology to the Selection of Refuse Disposal Sites," *Ground Water,* Vol. 7, No. 6, pp 9-13, 1969.

46. Hughes, G. M., et al, *Hydrogeology of Solid Waste Disposal Sites In Northeastern Illinois,* U.S. Environmental Protection Agency, Report SW–12D, 154 pp, 1971.

47. Hughes, G. M., et al, *Summary of Findings on Solid Waste Disposal Sites in Northeastern Illinois,* Illinois State Geological Survey Environmental Geology Note 45, 25 pp, 1971.

48. Cartwright, K., and M. R. McComas, "Geophysical Surveys in the Vicinity of Sanitary Landfills in Northeastern Illinois," *Ground Water,* Vol. 6, No. 5, pp 23–30, 1968.

49. Cartwright, K., and F. B. Sherman, *Evaluating Sanitary Landfill Sites in Illnois,* Illinois State Geological Survey Environmental Geology Note No. 27, 15 pp, August 1969.

50. Farvolden, R. N., and G. M. Hughes, "Sanitary Landfill Design," *Industrial Water Engineering,* Vol. 6, No. 8, pp 26–28, August 1969.

51. Andersen, J. R., and J. N. Dornbush, *Investigation of the Influence of Waste Disposal Practices on Ground Water Qualities,* Technical Completion Report, Water Resources Institute, South Dakota State Univ., 41 pp, Nov. 1968.

52. Andersen, J. R., and J. N. Dornbush, "Intercepting Trench at Landfill Renders Leachate Into Drinkable Source of Water," *Solid Wastes Management,* Vol. 15, No. 1, pp 60–68, 1972.

53. Andersen, J. R., *Studies of the Influence of Lagoons and Landfills on Groundwater Quality,* Water Resources Inst., South Dakota State Univ., Brookings, 47 pp, Dec. 1972. (NTIS: PB–214 138)

54. Weaver, L., "Refuse Disposal – Its Significance," *Ground Water,* Vol. 2, No. 1, pp 26-30, 1964.

55. Lane, B. E., and R. R. Parizek, "Leachate Movement in the Sub-Soil Beneath a Sanitary Landfill Trench Traced by Means of Suction Lysimeters," *Proceedings of the 2nd Mid-Atlantic Industrial Waste Conference,* Drexel Institute of Technology, pp 261-277, 1968.

56. Apgar, M. A., and D. Langmuir, "Ground-Water Pollution Potential of a Landfill Above the Water Table," *Ground Water,* Vol. 9, No. 6, pp 77-94, 1971.

57. Remson, I., et al, "Water Movement in an Unsaturated Sanitary Landfill," *Jour. Sanitary Engineering Div.,* Amer. Soc. of Civil Engineers, Vol. 94, No. SA 2, Paper 5904, pp 307-317, 1968.

58. Dair, F. R., "Seepage and Seepage Control Problems in Sanitary Landfills," *Agricultural Research Service ARS 41-147,* U.S. Dept. of Agriculture, pp 14-16, 1969.

59. Qasim, S. R., and J. C. Burchinal, "Leaching from Simulated Landfills," *Jour. Water Pollution Control Fed.,* Vol. 42, No. 3, Part 1, pp 371-379, 1970.

60. Kaufmann, R. F., *Hydrogeology of Solid Waste Disposal Sites in Madison, Wisconsin,* Technical Report, Univ. of Wisconsin, Madison, Wisconsin Water Resources Center, 361 pp, 1970. (NTIS: PB-196 360)

61. Schneider, W. J., *Hydrologic Implications of Solid-Waste Disposal,* U.S. Geological Survey Circular 601-F, 10 pp, 1970.

62. Williams, R. E., and A. T. Wallace, *Hydrogeological Aspects of the Selection of Refuse Disposal Sites in Idaho,* Pamphlet 145, Idaho Bureau of Mines and Geology, Moscow, Idaho, 31 pp, 1970.

63. Hughes, G., et al, *Pollution of Groundwater Due to Municipal Dumps*, Canada Dept. of Energy, Mines and Resources, Inland Waters Branch, Ottawa, Technical Bulletin No. 42, 98 pp, 1971.

64. Emery, W. T., *Sanitary Landfill Leachate Travel in Various Soil Media − A Bibliography*, Technical Information Center, College of Technology, Univ. of Vermont, Burlington, 11 pp, Dec. 1971.

65. Salvato, J. A., et al, "Sanitary Landfill − Leaching Prevention and Control," *Jour. Water Pollution Control Fed.*, Vol. 43, No. 10, pp 2084-2100, 1971.

66. Sheffer, H. W., et al, *Case Studies of Municipal Waste Disposal Systems*, Information Circular 8498, Bureau of Mines, Pittsburgh, Pa., Eastern Field Operation Center, 36 pp, 1971.

67. Fungaroli, A. A., *Pollution of Subsurface Water by Sanitary Landfills, Vol. 1*, U.S. Environmental Protection Agency Report EPA–SW–12RG.-71, 198 pp, 1971.
    (NTIS: PB-209 000)

68. Fungaroli, A. A., *Pollution of Subsurface Water by Sanitary Landfills, Vol. 2*, U.S. Environmental Protection Agency Report EPA–SW–12RG. 1-71, 216 pp, 1971.
    (NTIS: PB-209 001)

69. Fungaroli, A. A., *Pollution of Subsurface Water by Sanitary Landfills, Vol. 3*, U.S. Environmental Pretection Agency Report EPA–SW–12RG. 2-71, 169 pp, 1971.
    (NTIS: PB-209 002)

70. Zanoni, A. E., *Groundwater Pollution from Sanitary Landfills and Refuse Dump Grounds, A Critical Review*, Research Report 69, Wisconsin Dept. of Natural Resources, 43 pp, 1971.

71. Zanoni, A. E., "Ground Water Pollution and Sanitary Landfills − A Critical Review," *Ground Water*, Vol. 10, No. 1, pp 3-16, 1972.

72. Seitz, H. R., "Investigation of a Landfill in Granite-Loess Terrane," *Ground Water*, Vol. 10, No. 4, pp 35-41, 1972.

73. Otton, E. G., *Solid-Waste Disposal in the Geohydrologic Environment of Maryland*, Maryland Geological Survey Report of Investigations No. 18, 59 pp, 1972.

74. Anon., "Sanitary Landfills; The Latest Thinking," *Civil Engineering,* Vol. 43, No. 3, pp 69-71, 1973.

75. Hanes, R. E., et al, *Effects of Deicing Salts on Water Quality and Biota,* National Cooperative Highway Research Program Report 91, 70 pp, 1970.

76. Walker, W. H., "Limiting Highway Salt Pollution of Area Water Supplies," *Rural and Urban Roads,* pp 74-75, March 1971.

77. Struzeski, E. J., Jr., "Environmental Impact of Highway Deicing," *Street Salting – Urban Water Quality Workshop Proc.,* State College of Forestry, Syracuse, N. Y., pp 14-19, July 1971.

78. Field, R., et al, *Water Pollution and Associated Effects from Street Salting,* EPA–R2–73–257, U.S. Environmental Protection Agency, Cincinnati, Ohio, 48 pp, May 1973.

79. Walker, W. H., "Salt Piling – A Source of Water Supply Pollution," *Pollution Engineering,* Vol. 1, No. 4, pp 30-33, 1970.

80. Hutchinson, F. E., "Environmental Pollution from Highway Deicing Compounds," *Jour. Soil and Water Conservation,* Vol. 25, No. 4, pp 144-146, 1970.

81. Broecker, W. S., et al, "Road Salt as an Urban Tracer," *Street Salting – Urban Water Quality Workshop Proceedings,* State College of Forestry, Syracuse, N. Y., pp 24-38, July 1971. (NTIS: Conf-710531-1)

82. Coogan, G. J., "The Increase in Chlorides Experienced in Massachusetts Water Supplies," *New England Water Works Assoc. Jour.,* Vol. 85, No. 2, pp 173-178, June 1971.

83. Huling, E. E., and T. C. Hollocher, "Groundwater Contamination by Road Salt – Steady-State Concentrations in East Central Massachusetts," *Science,* Vol. 176, No. 4032, pp 288-290, 1972.

84. Dennis, H. W., "Salt Pollution of a Shallow Aquifer – Indianapolis, Indiana," *Ground Water,* Vol. 11, No. 4, pp 18-22, 1973.

85. Woodward, F. L., et al, "Experiences with Ground Water Contamination in Unsewered Areas in Minnesota," *Amer. Jour. of Public Health,* Vol. 51, No. 8, pp 1130-1136, 1961.

86. Polta, R. C., "Septic Tank Effluents," *Water Pollution by Nutrients – Sources, Effects, and Controls,* Water Resources Research Center, Univ. of Minnesota, Minneapolis, WRRC Bulletin 13, pp 53-57, June 1969.

87. Hall, M. W., *Water Quality Degradation by Septic Tank Drainage,* Project Termination Report, Water Resources Research Center, Univ. of Maine, Orono, 9 pp, June 1970.
(NTIS: PB-195 307)

88. Patterson, J. W., et al, *Septic Tanks and the Environment,* Final Report, Illinois Inst. for Environmental Quality, Chicago, 105 pp, June 1971.
(NTIS: PB-204 519)

89. Crosby, J. W., III, et al, *Investigation of Techniques to Provide Advance Warning of Ground-Water Pollution Hazards with Special Reference to Aquifers in Glacial Outwash,* Final Report, Water Research Center, Washington State Univ., Pullman, 148 pp, Aug. 1971.
(NTIS: PB-203 748)

90. Waltz, J. P., "Methods of Geologic Evaluation of Pollution Potential at Mountain Homesites," *Ground Water,* Vol. 10, No. 1, pp 42-47, 1972.

91. Baker, E. T., Jr., and J. Rawson, *Ground-Water Pollution in the Vicinity of Toledo Bend Reservoir, Texas,* Progress Report, 1972, U.S. Geological Survey Open-File Report, 24 pp, Aug. 1972.

92. Anon., "Findings and Recommendations on Underground Waste Disposal," *Jour. Amer. Water Works Assoc.,* Vol. 45, No. 12, pp 1295-1297, 1953.

93. Ives, R. E., and G. E. Eddy, *Subsurface Disposal of Industrial Wastes,* Interstate Oil Compact Commission Study, Oklahoma City, 109 pp, June 1968.

94. Ulrich, A. A., "Chloride Contamination of Ground Water in Ohio," *Jour. Amer. Water Works Assoc.,* Vol. 47, No. 2, pp 151-152, 1955.

95. Parks, W. W., "Decontamination of Ground Water at Indian Hill," *Jour. Amer. Water Works Assoc.,* Vol. 51, No. 5, pp 644-646, 1959.

96. Petri, L. R., "The Movement of Saline Ground Water in the Vicinity of Derby, Colorado," *Soc. for Water Treatment and Examination, Proc.,* Vol. 11, Pt. 2, pp 88-93, 1962.

97. Walker, T. R., "Ground-Water Contamination in the Rocky Mountain Arsenal Area, Denver, Colorado," *Geological Soc. of Amer. Bulletin,* Vol. 72, No. 3, pp 489-494, 1961.

98. Swenson, H. A., "The Montebello Incident," *Soc. for Water Treatment and Examination, Proc.,* Vol. 11, Pt. 2, pp 84-88, 1962.

99. Evans, R., "Industrial Wastes and Water Supplies," *Jour. Amer. Water Works Assoc.,* Vol. 57, No. 4, pp 625-628, 1965.

100. California Dept. of Water Resources, *Fresno-Clovis Metropolitan Area Water Quality Investigation,* Bulletin 143-3, 27 pp, 1965.

101. Price, D., "Rate and Extent of Migration of a 'One-Shot' Contaminant in an Alluvial Aquifer in Keizer, Oregon," *Geological Survey Research 1967,* U.S. Geological Survey Prof. Paper 575–B, pp B217-B220, 1967.

102. Bergstrom, R. E., "Hydrogeological Studies are Key to Safety in Waste Management Programs," *Water and Sewage Works,* Vol. 116, No. 4, pp 149-155, 1969.

103. Maehler, C. Z., and A. E. Greenberg, "Identification of Petroleum Industry Wastes in Groundwaters," *Jour. Water Pollution Control Fed.,* Vol. 34, No. 12, pp 1262-1267, 1962.

104. Deutsch, M., "Phenol Contamination of an Artesian Glacial-Drift Aquifer at Alma, Michigan, U. S. A.," *Soc. for Water Treatment and Examination, Proc.,* Vol. 11, Pt. 2, pp 94-100, 1962.

105. Zimmerman, W., "Pollution of Water and Soil by Miscellaneous Petroleum Products," *Sixth Congress, Intl. Water Supply Assoc.,* pp B1-80, 1964.

106. Grubb, H. F., "Effects of a Concentrated Acid on Water Chemistry and Water Use in a Pleistocene Outwash Aquifer," *Ground Water,* Vol. 8, No. 5, pp 4-7, 1970.

107. Van der Warden, M., et al, "Transport of Mineral Oil Components to Groundwater — 1. Model Experiments on the Transfer of

Hydrocarbons from a Residual Oil Zone to Trickling Water," *Water Research,* Vol. 5, No. 5, pp 213-226, 1971.

108. Williams, D. E., and D. G. Wilder, "Gasoline Pollution of a Ground Water Reservoir — A Case History," *Ground Water,* Vol. 9, No. 6, pp 50-54, 1971.

109. McKee, J. E., et al, "Gasoline in Groundwater," *Jour. Water Pollution Control Fed.,* Vol. 44, No. 2, pp 293-302, 1972.

110. Matis, J. R., "Petroleum Contamination of Ground Water in Maryland," *Ground Water,* Vol. 9, No. 6, pp 57-61, 1971.

111. Collins, A. G., "Oil and Gas Wells — Potential Polluters of the Environment?" *Jour. Water Pollution Control Fed.,* Vol. 43, No. 12, pp 2383-2393, 1971.

112. Committee on Environmental Affairs, *The Migration of Petroleum Products in Soil and Ground Water — Principles and Countermeasures,* Publication No. 4149, Amer. Petroleum Inst., Washington, D. C., 36 pp, Dec. 1972.

113. Davids, H. W., and M. Lieber, "Underground Water Contamination by Chromium Wastes," *Water and Sewage Works.* Vol. 98, No. 12, pp 528-534, 1951.

114. Lieber, M., and W. F. Welsh, "Contamination of Ground Water by Cadmium," *Jour. Amer. Water Works Assoc.,* Vol. 46, No. 6, pp 541-547, 1954.

115. Lieber, M., et al, "Cadmium and Hexavalent Chromium in Nassau County Ground Water," *Jour. Amer. Water Works Assoc.,* Vol. 56, pp 739-747, 1964.

116. Perlmutter, N. M., et al, "Movement of Waterborne Cadmium and Hexavalent Chromium Wastes in South Farmingdale, Nassau County, Long Island, New York," *Short Papers in Geology and Hydrology,* U.S. Geological Survey Prof. Paper 475-C, pp C179-C184, 1963.

117. Perlmutter, N. M., and M. Lieber, *Dispersal of Plating Wastes and Sewage Contaminants in Groundwater and Surface Water, South Farmingdale–Massapequa Area, Nassau County, New York,* U.S. Geological Survey Water Supply Paper 1879-G, 67 pp, 1970.

118. Rice and Co., *Engineering Economic Study of Mine Drainage Control Techniques,* Appalachian Regional Commission Report on Acid Mine Drainage in Appalachia, Appendix B, Pittsburgh, Pa., 281 pp, 1969.

119. Emrich, G. H., and G. L. Merritt, "Effects of Mine Drainage on Ground Water," *Ground Water,* Vol. 7, No. 3, pp 27-32, 1969.

120. Moulder, E. A., "Water Pollution Control Creates Demand for Groundwater Hydrologists," *Mining Engineering,* pp 90-91, Feb. 1970.

121. Ahmad, M. U., "A Hydrological Approach to Control Acid Mine Pollution for Lake Hope," *Ground Water,* Vol. 8, No. 6, pp 19-24, 1970.

122. Ahmad, M., *Proceedings of Acid Mine Drainage Workshop,* Ohio Univ., Athens, Ohio, 167 pp, Aug. 1971.

123. Mink, L. L., et al, *Effect of Industrial and Domestic Effluents on the Water Quality of the Coeur d'Alene River Basin,* Idaho Bureau of Mines and Geology Pamphlet 149, 30 pp, 1971.

124. Mink, L. L., et al, "Effect of Early Day Mining Operations on Present Day Water Quality," *Ground Water,* Vol. 10, No. 1, pp 17-26, 1972.

125. Galbraith, J. H., et al, "Migration and Leaching of Metals from Old Mine Tailings Deposits," *Ground Water,* Vol. 10, No. 3, pp 33-44, 1972.

126. Merkel, R. H., "The Use of Resistivity Techniques to Delineate Acid Mine Drainage in Ground Water," *Ground Water,* Vol. 10, No. 5, pp 38-42, 1972.

127. Subcommittee on Water Problems Associated with Oil Production in the United States, *Water Problems Associated with Oil Production in the United States,* Interstate Oil Compact Commission, 88 pp, June 1967.

128. Petroleum Equipment and Services,"Salt Water Disposal and Oilfield Water Conservation," *Petroleum Equipment and Services,* Vol. 30, No. 4, pp 22, 24-26, 28, 1967.

129. Anon., "Crackdown on Oil Field Pollution," *Petroleum Engineer,* Vol. 39, No. 7, pp 33-36, 1967.

130. Crain, J., *Ground-Water Pollution from Natural Gas and Oil Production in New York,* New York State Water Resources Commission Report of Investigation No. RI-5, 15 pp, 1969.

131. Bain, G. L., *Salty Groundwater in the Pocatalico River Basin,* West Virginia Geological and Economic Survey Circular Series, No. 11, 31 pp, Oct. 1970.

132. Boster, R. S., *A Study of Ground-Water Contamination Due to Oil-Field Brines in Morrow and Delaware Counties, Ohio, with Emphasis on Detection Utilizing Electrical Resistivity Techniques,* Water Resources Research Center, Ohio State Univ., Columbus, 191 pp, 1967.

133. Lehr, J. H., *A Study of Groundwater Contamination Due to Saline Water Disposal in the Morrow County Oil Fields,* Research Project Completion Report, Water Resources Center, Ohio State Univ., Columbus, 81 pp, March 1969.

134. Pettyjohn, W. A., "Water Pollution by Oil-Field Brines and Related Industrial Wastes in Ohio," *Ohio Jour. of Science,* Vol. 71, No. 5, pp 257-269, 1971.

135. Eddy, G. E., "The Effectiveness of Michigan's Oil and Gas Conservation Law in Preventing Pollution of the State's Ground Waters," *Ground Water,* Vol. 3, No. 2, pp 35-36, 1965.

136. Krieger, R. A., and G. E. Hendrickson, *Effects of Greensburg Oilfield Brines on the Streams, Wells, and Springs of the Upper Green River Basin, Kentucky,* Kentucky Geological Survey Report of Investigations 2, Ser. X, 36 pp, 1960.

137. Hopkins, H. T., *The Effect of Oilfield Brines on the Potable Ground Water in the Upper Big Pitman Creek Basin, Kentucky,* Kentucky Geological Survey Report of Investigations 4, Ser. X, 36 pp, 1963.

138. Wait, R. L., and M. J. McCollum, "Contamination of Fresh Water Aquifers Through an Unplugged Oil-Test Well in Glynn County, Georgia," *Georgia Geological Survey Mineral Newsletter,* Vol. 16, No. 3-4, pp 74-80, 1963.

139. Fryberger, J. S., *Rehabilitation of a Brine-Polluted Aquifer,* EPA–R2–72–014, Office of Research and Monitoring, U. S. Environmental Protection Agency, Washington, D. C., 61 pp, Dec. 1972.

140. Powell, W. J., et al, *Water Problems Associated with Oil Production in Alabama,* Alabama Geological Survey Circular 22, 63 pp, 1963.

141. Knowles, D. B., "Hydrologic Aspects of the Disposal of Oil-Field Brines in Alabama," *Ground Water,* Vol. 3, No. 2, pp 22-27, 1965.

142. Irwin, J. H., and R. B. Morton, *Hydrogeologic Information on the Glorieta Sandstone and the Ogallala Formation in the Oklahoma Panhandle and Adjoining Areas as Related to Underground Waste Disposal,* U.S. Geological Survey Circular 630, 26 pp, 1969.

143. McMillion, L. G., "Hydrologic Aspects of Disposal of Oil-Field Brines in Texas," *Ground Water,* Vol. 3, No. 4, pp 36-42, 1965.

144. Payne, R. D., "Saltwater Pollution Problems in Texas," *Jour. Petroleum Technology,* Vol. 18, pp 1401-1407, Oct. 1966.

145. Burke, R. G., "Texas Toughens Anti-Pollution Line," *Oil and Gas Jour.,* Vol. 64, No. 1, pp 47-48, 1966.

146. Page, R. D., "Pollution Control for Oil-Field Brines," *Drill Bit,* Vol. 15, No. 9, pp 32-36, 1967.

147. Burnitt, S. C., et al, *Reconnaissance Survey of Salt Water Disposal in the Mexia, Negro Creek, and Cedar Creek Oil Fields, Limestone County, Texas,* Texas Water Comm. Memo. Report 62-02, 27 pp, 1962.

148. Burnitt, S. C., *Reconnaissance of Soil Damage and Ground-Water Quality, Fisher County, Texas,* Texas Water Comm. Memo. Report 63-02, 49 pp, 1963.

149. Fink, B. E., *Investigation of Ground- and Surface-Water Contamination Near Harrold, Willbarger County, Texas,* Texas Water Comm. Report LD–0365, 23 pp, 1965.

150. Preston, R. D., *Occurrence and Quality of Groundwater in Shackelford County, Texas,* Texas Water Development Board Report 100, 48 pp, Oct. 1969.

151. Rice, I. M., "Salt Water Disposal in the Permian Basin," *Producers Monthly,* Vol. 32, No. 3, pp 28-30, 1968.

152. McMillion, L. G., "Ground Water Reclamation by Selective Pumping," *Trans. of AIME,* Vol. 250, pp 11-15, March 1971.

153. Rold, J. W., "Pollution Problems in the 'Oil Patch'," *Amer. Assoc. Petroleum Geologists Bulletin,* Vol. 55, No. 6, pp 807-809, 1971.

154. Williams, C. C., and C. K. Bayne, "Ground-Water Conditions in Elm Creek Valley, Barber County, Kansas, with Special Reference to Contamination of Ground Water by Oil Field Brine," *Kansas State Geological Survey Bulletin 64,* 1946 Reports of Studies, Pt. 3, pp 77-124, 1946.

155. Jones, O. S., *Fresh Water Protection from Pollution Arising in the Oil Fields,* Univ. of Kansas Press, Lawrence, 132 pp, 1950.

156. Latta, B. F., "Fresh Water Pollution Hazards Related to the Petroleum Industry in Kansas," *Kansas Academy of Science Trans.,* Vol. 66, No. 1, pp 25-33, 1963.

157. California Div. of Water Resources, *Oil Field Waste Water Disposal, Raisin City Oil Field, Fresno County,* Water Quality Investigations, California Dept. of Public Works, 26 pp, March 1955.

158. Anon., "California Acts to Cut Water Pollution," *Oil and Gas Jour.,* Vol. 55, No. 28, pp 66-67, 1957.

159. Harmeson, R. H., and O. W. Vogel, "Artificial Recharge and Pollution of Ground Water," *Ground Water,* Vol. 1, No. 1, pp 11-15, 1963.

160. Wichman, S. H., and G. K. Ahlers, "Recharging Conserves Nuclear Reactor Cooling Water," *Water and Wastes Engineering,* pp 79-81, March 1967.

161. Preul, H. C., "Contaminants in Groundwaters Near Waste Stabilization Ponds," *Jour. Water Pollution Control Fed.,* Vol. 40, No. 4, pp 649-669, 1968.

162. Middletown, F. M., and R. L. Bunch, "Challenge for Waste Water Lagoons," *2nd International Symposium for Waste Treatment Lagoons,* Kansas City, Missouri, pp 364-366, June 1970.

163. Tossey, D., "Digested Sludge Lagoons," *2nd International Symposium for Waste Treatment Lagoons*, Kansas City, Missouri, pp 333-334, June 1970.

164. Hackbarth, D. A., "Field Study of Subsurface Spent Sulfite Liquor Movement Using Earth Resistivity Measurements," *Ground Water*, Vol. 9, No. 3, pp 11-16, 1971.

165. Wells, D. M., *The Effect of Unlined Treated Sewage Storage Ponds on Water Quality in the Ogallala Formation*, Project Completion Report, Water Resources Center, Texas Tech. Univ., Lubbock, 36 pp, Dec. 1971.
(NTIS: PB-206 517)

166. Leggat, E. R., et al, *Liquid-Waste Disposal at the Linfield Disposal Site, Dallas, Texas*, U.S. Geological Survey Report, 28 pp, May 1972.

167. de Laguna, W., and J. O. Blomeke, *The Disposal of Power Reactor Waste into Deep Wells*, Atomic Energy Commission Report ORNL–CF–57–6–23, U. S. Atomic Energy Commission Office of Technical Information, 48 pp, June 1957.

168. Roedder, E., *Problems in the Disposal of Acid Aluminum Nitrate High-Level Radioactive Waste Solutions by Injection into Deep-Lying Permeable Formations*, U. S. Geological Survey Bulletin 1088, 65 pp, 1959.

169. Kaufman, W. J., et al, "Disposal of Radioactive Wastes into Deep Geologic Formations," *Jour. Water Pollution Control Fed.*, Vol. 33, No. 1, pp 73-84, 1961.

170. Belter, W. G., "Waste Management Activities in the Atomic Energy Commission," *Ground Water*, Vol. 1, No. 1, pp 17-24, 1963.

171. Mawson, C. A., *Management of Radioactive Wastes*, D. Van Nostrand Co., Princeton, N. J., 196 pp, 1965.

172. Clebsch, A., Jr., and E. H. Baltz, "Progress in the United States of America Toward Deep-Well Disposal of Liquid and Gaseous Radioactive Wastes," *Symposium of Intl. Atomic Energy Agency, Vienna, and European Nuclear Energy Agency*, pp 591-605, 1967.

173. Lynch, E. J., *Transport of Radionuclides by Groundwater – Some Theoretical Aspects,* Report HNS–1229–61, Hazelton Nuclear Science Corp., Palo Alto, Calif., and Corrol E. Bradberry and Associates, Los Altos, Calif., 56 pp, Nov, 1964. (NTIS)

174. Stead, F. W., "Distribution in Groundwater of Radionuclides from Underground Nuclear Explosions," *Engineering with Nuclear Explosives – Proceedings of Third Plowshare Symposium, April 1964,* U.S. Atomic Energy Commission Div. of Technical Information Report TID–7695, pp 127-138, 1964.

175. Champlin, J. B. F., and G. G. Eichholz, "The Movement of Radioactive Sodium and Ruthenium Through a Simulated Aquifer," *Water Resources Research,* Vol. 4, No. 1, pp 147-158, 1968.

176. Witkowski, E. J., and J. F. Manneschmidt, "Ground Disposal of Liquid Wastes at Oak Ridge National Laboratory," *Proceedings of 2nd Ground Disposal of Radioactive Wastes Conf.,* U.S. Atomic Energy Commission Div. of Technical Information Report TID–7628, Book 2, pp 506-512, 1962.

177. Reichert, S. O., "Radionuclides in Groundwater at the Savannah River Plant Waste Disposal Facilities," *Jour. Geophysical Research,* Vol. 67, No. 11, pp 4363-4374, 1962.

178. Reichert, S. O., and J. W. Fenimore, "Lithology and Hydrology of Radioactive Waste-Disposal Sites, Savannah River Plant, South Carolina," *Geologic Soc. of Amer. Engineering Geology Case Histories,* No. 5, pp 53-69, 1964.

179. Proctor, J. F., and I. W. Marine, "Geologic, Hydrologic, and Safety Considerations in the Storage of Radioactive Wastes in a Vault Excavated in Crystalline Rock," *Nuclear Science Engineering,* Vol. 22, No. 3, pp 350-365, 1965.

180. Gardner, M. C., and C. E. Downs, *Evaluation of the Project Dribble Site, Hattiesburg, Mississippi, for Disposition Including Identification of Restrictions,* NTIS: NVO–1229–145 (Pt. 1), 49 pp, March 1971.

181. Lynn, R. D., and Z. E. Arlin, "Deep Well Construction for the Disposal of Uranium Mill Tailing Water by The Anaconda Company at Grants, New Mexico," *Soc. of Mining Engineers Trans.,* Vol. 223, No. 3, pp 230-237, 1962.

182. Purtymun, W. D., et al, "Distribution of Radioactivity in the Alluvium of a Disposal Area at Los Alamos, New Mexico," *Geological Survey Research 1966,* U.S. Geological Survey Prof. Paper 550–D, pp D250-D252, 1966.

183. Peckham, A. E., "Underground Waste Disposal Studies — Chemical Processing Plant Area," *Proceedings of Ground Disposal of Radioactive Wastes Conf.,* Kaufman, W. J., Ed., U. S. Atomic Energy Commission Div. of Technical Information Report TID-7621, pp 157-166, 1961.

184. Jones, P. H., and E. Shuter, "Hydrology of Radioactive Waste Disposal in the MTR–ETR Area, National Reactor Testing Station, Idaho," *Short Papers in Geology and Hydrology,* U.S. Geological Survey Prof. Paper 450–C, pp C113-C116, 1962.

185. Morris, D. A., et al, *Hydrology of Subsurface Waste Disposal, National Reactor Testing Station, Idaho — Annual Progress Report,* 1963, U. S. Geological Survey Open-File Report (IDO-22046), 228 pp, 1964.

186. Schoen, R., "Hydrochemical Study of the National Reactor Testing Station, Idaho," *24th International Geological Congress, Section 11, Hydrogeology,* Montreal, pp 306-314, 1972.

187. Nebeker, R. L., and L. T. Lakey, "Liquid Waste Management at the NRTS Test Reactor Area," *Nuclear Engineering,* Part 23, Vol. 68, No. 123, pp 10-17, 1972.

188. Brown, R. E., et al, "Geological and Hydrological Aspects of the Disposal of Liquid Radioactive Wastes," *Proceedings of Seminar on Sanitary Engineering Aspects,* Atomic Energy Commission Div. of Technical Information Report TID-7517, Part 16, pp 413-425, 1956.

189. Raymond, J. R., and W. H. Bierschenk, "Hydrologic Investigations at Hanford," *Amer. Geophysical Union Trans.,* Vol 38, No. 5, pp 724-729, 1957.

190. Bierschenk, W. H., "Hydrologic Aspects of Radioactive Waste Disposal," *Jour. Sanitary Engineering Div.,* Amer. Soc. of Civil Engineers, Vol. 84, No. SA6, Paper 1835, 11 pp, 1958.

191. Bierschenk, W. H., *Aquifer Characteristics and Ground-Water Movement at Hanford (Wash.)*, U.S. Atomic Energy Commission Div. of Technical Information Report TID-4500, 77 pp, 1959.

192. Brown, R. E., and J. R. Raymond, "The Measurement of Hanford's Geohydrologic Features Affecting Waste Disposal," *Proceedings of 2d Ground Disposal of Radioactive Wastes Conf.*, Book 1, Morgan, J. M., Jr., et al, Eds., U.S. Atomic Energy Commission Div. of Technical Information Report TID-7628, pp 77-79, 1962.

193. Brown, D. J., and J. R. Raymond, "Radiologic Monitoring of Ground Water at the Hanford Project," *Jour. Amer. Water Works Assoc.*, Vol. 54, No. 10, pp 1201-1212, 1962.

194. Raymond, J. R., and V. L. McGhan, *Scintillation Probe Results on 200 Area Waste Disposal Site Monitoring Wells*, Hanford Atomic Products Operation Report HW-84577, General Electric Co., Richland, Wash., 62 pp, 1965.
    (NTIS)

195. Brown, D. J., "Migration Characteristics of Radionuclides Through Sediments Underlying the Hanford Reservation," *Disposal of Radioactive Wastes into the Ground*, Intl. Atomic Energy Agency (STI/PUB/156), Vienna, pp 215-225, 1967.

196. Essig, T. H., *Radiological Status of the Ground-Water Beneath the Hanford Project, January-June, 1968*, Battelle Memorial Inst. Pacific Northwest Lab. Report BNWL-984, Richland, Wash., 20 pp, Jan. 1969.
    (NTIS: BNWL-984)

197. Benham, D. H., *Radiological Status of the Groundwater Beneath the Hanford Project, January-June 1969*, Battelle-Northwest, Richland, Wash., 20 pp, Nov. 1969.
    (NTIS: BNWL-1233)

198. Benham, D. E., *Radiological Status of the Groundwater Beneath the Hanford Project, July-December 1969*, Battelle-Northwest, Richland, Wash., 20 pp, May 1970.
    (NTIS: BNWL-1392)

199. Essig, T. H., *Radiological Status of the Groundwater Beneath the Hanford Project, January-June, 1970*, National Technical Information Service: BNWL-1539, 23 pp, Dec. 1970.

200. Essig, T. H., *Radiological Status of the Groundwater Beneath the Hanford Project, July-December, 1970,* National Technical Information Service: BNWL-1613, 20 pp, Sept. 1971.

201. Kipp, K. L., Jr., *Radiological Status of the Groundwater Beneath the Hanford Project, January-June, 1971,* Battelle-Pacific Northwest Lab. Report No. BNWL-1649, Richland, Wash., 24 pp, April 1972.
(NTIS: BNWL-1649)

202. LaSala, A. M., and G. C. Doty, *Preliminary Evaluation of Hydrologic Factors Related to Radioactive Waste Storage in Basaltic Rocks at the Hanford Reservation, Washington,* U.S. Geological Survey Open-File Report, and Atomic Energy Commission Report TID-25764, 68 pp, 1971.

203. Stewart, B. A., et al, *Distribution of Nitrates and Other Water Pollutants Under Fields and Corrals in the Middle South Platte Valley of Colorado,* U.S. Dept. of Agriculture, Agricultural Research Service, ARS 41-134, 206 pp, Dec. 1967.

204. Stewart, B. A., et al, "Agriculture's Effect on Nitrate Pollution of Groundwater," *Jour. Soil and Water Conservation,* Vol. 23, No. 1, pp 13-15, 1968.

205. Robbins, J. W. D., and G. J. Kriz, "Relation of Agriculture to Groundwater Pollution: A Review," *Trans, Amer. Soc. of Agricultural Engineers,* Vol. 12, No. 3, pp 397-403, 1969.

206. Smith, G. E., *Contribution of Fertilizers to Water Pollution,* Paper No. 7, 2nd Compendium of Animal Waste Management, 16 pp, June 1969.

207. Biggar, J. W., and R. B. Corey, "Agricultural Drainage and Eutrophication, " *Eutrophication: Causes, Consequences, Correctives,* National Academy of Sciences, Washington, D. C., pp 404-445, 1969.

208. Moore, T. M., *Water Geochemistry, Hog Creek Basin, Central Texas,* Baylor Geological Studies Bulletin No. 18, Dept. of Geology, Baylor Univ., Waco, 44 pp, Spring 1970.

209. LeGrand, H. E., "Movement of Agricultural Pollutants with Groundwater," *Agricultural Practices and Water Quality,* Willrich and Smith (Eds.), Iowa State Univ. Press, Ames, pp 303-313, 1970.

210. Viets, F. G., Jr., "Water Quality in Relation to Farm Use of Fertilizer," *Bioscience,* Vol. 21, No. 10, pp 460-467, 1971.

211. Resnik, A. V., and J. M. Rademacher, *Animal Waste Runoff — A Major Water Quality Challenge,* Paper No. 1, 2nd Compendium of Animal Waste Management, 21 pp, June 1969.

212. Resnik, A. V., and J. M. Rademacher, "Animal Waste Runoff — A Major Water Quality Challenge," *Water Quality Management Problems in Arid Regions,* Federal Water Quality Admin. Water Pollution Control Research Series 13030 DYY 6/69, pp 95-105, Oct. 1970.

213. Rademacher, J. M., and A. V. Resnik, "Feedlot Pollution Control — A Profile for Action," *Animal Waste Management,* Proceedings of Cornell Univ. Conf. on Agricultural Waste Management, pp 193-202, 1969.

214. Miner, J. R., and T. L. Willrich, "Livestock Operations and Field-Spread Manure as Sources of Pollutants," *Agricultural Practices and Water Quality,* Willrich and Smith (Eds.), Iowa State Univ. Press, Ames, pp 231-240, 1970.

215. McCalla, T. M., et al, "Manure Decomposition and Fate of Breakdown Products in Soil," *Agricultural Practices and Water Quality,* Willrich and Smith (Eds.), Iowa State Univ. Press, Ames, pp 241-255, 1970.

216. Loehr, R. C., "Effluent Quality from Anaerobic Lagoons Treating Feedlot Wastes," *Jour. Water Pollution Control Fed.,* Vol. 39, No. 3, pp 384-391, 1967.

217. Webber, L. R., and T. H. Lane, "The Nitrogen Problem in the Land Disposal of Liquid Manures," *Animal Waste Management,* Proceedings of Cornell Univ. Conf. on Agricultural Waste Management, pp 124-130, 1969.

218. Overman, A. R., et al, "Land Disposal of Dairy Farm Waste," *Relationship of Agriculture to Soil and Water Pollution,* Proceedings of Cornell Univ. Conf. on Agricultural Waste Management, pp 123-126, Jan. 1970.

219. Loehr, R. C., "Drainage and Pollution from Beef Cattle Feedlots," *Jour. Sanitary Engineering Div.,* Amer. Soc. of Civil Engineers, Vol. 96, No. SA6, pp 1295-1309, 1970.

220. Fogg, C. E., "Livestock Waste Management and the Conservation Plan," *Livestock Waste Management and Pollution Abatement,* Proceedings of Intl. Symposium on Livestock Wastes, Amer. Soc. of Agricultural Engineers, pp 34-35, April 1971.

221. Concannon, T. J., and E. J. Genetelli, "Ground-Water Pollution Due to High Organic Manure Loadings," *Livestock Waste Management and Pollution Abatement,* Proceedings of Intl. Symposium on Livestock Wastes, Ohio State Univ., pp 249-253, April 1971.

222. Viets, F. G., "Cattle Feedlot Pollution," *Animal Waste Management,* Proceedings of National Symposium on Animal Waste Management, The Airlie House, Warrenton, Virginia, pp 97-105, Sept. 1971.

223. Stewart, B. A., et al, "Nitrate and Other Water Pollutants under Fields and Feedlots," *Environmental Science and Technology,* Vol. 1, No. 9, pp 736-739, 1967.

224. Evans, C. E., "Research on Abatement of Pollution and Management of Organic Wastes from Cattle Feedlots in Northeastern Colorado and Eastern Nebraska," *Proceedings of Animal Waste Management Conf.,* Kansas City, Missouri, pp 20-22, Feb. 1969.

225. Mielke, L. N., et al, "Groundwater Quality and Fluctuations in a Shallow Unconfined Aquifer under a Level Feedlot," *Relationship of Agriculture to Soil and Water Pollution,* Proceedings of Cornell Univ. Conf. on Agricultural Waste Management, pp 31-40, 1970.

226. Gilbertson, C. B., et al, "Runoff, Solid Wastes, and Nitrate Movement on Beef Feedlots," *Jour. Water Pollution Control Fed.,* Vol. 43, No. 3, Part 1, pp 483-493, 1971.

227. Lorimer, J. C., et al, "Nitrate Concentrations in Groundwater Beneath a Beef Cattle Feedlot," *Water Resources Bulletin,* Vol. 8, No. 5, pp 999-1005, 1972.

228. Rademacher, J. M., "Animal Waste Pollution – Overview of the Problem," *Proceedings of Animal Waste Management Conf.,* Kansas City, Missouri, pp 7-9, 1969.

229. Gillham, R. W., and L. R. Webber, "Groundwater Contamination," *Water and Pollution Control,* Vol. 106, No. 5, pp 54-57, 1968.

230. Gillham, R. W., and L. R. Webber, "Nitrogen Contamination of Groundwater by Barnyard Leachates," *Jour. Water Pollution Control Fed.*, Vol. 41, No. 10, pp 1752-1762, 1969.

231. Frink, C. R., "The Nitrogen Cycle of a Dairy Farm," *Relationship of Agriculture to Soil and Water Pollution*, Proceedings of Cornell Univ. Conf. on Agricultural Waste Management, pp 127-133, Jan. 1970.

232. Miller, W. D., *Infiltration Rates and Groundwater Quality Beneath Cattle, Texas High Plains*, Water Pollution Control Research Series Report, U.S. Environmental Protection Agency, Water Quality Office, EPA–WQP–16060–EGS–01/71, Jan. 1971.
(NTIS: PB-203 681)

233. Miller, W. D., "Subsurface Distribution of Nitrates Below Commercial Cattle Feedlots, Texas High Plains," *Water Resources Bulletin*, Vol. 7, No. 5, pp 941-950, 1971.

234. Crosby, J. W., III, et al, "Migration of Pollutants in a Glacial Outwash Environment, 2," *Water Resources Research*, Vol. 7, No. 1, pp 204-208, 1971.

235. Adriano, D. C., et al, "Fate of Inorganic Forms of N and Salt from Land Disposal Manures from Dairies," *Livestock Waste Management and Pollution Abatement*, Proceedings Intl. Symposium on Livestock Wastes, Amer. Soc. of Agricultural Engineers, pp 243-246, 1971.

236. Adriano, D. C., et al, "Nitrate and Salt in Soils and Ground Waters from Land Disposal of Dairy Manure," *Soil Science Soc. of Amer. Proceedings*, Vol. 35, pp 759-762, 1971.

237. Hutchinson, F. E., et al, *Effect of Animal Wastes Applied to Soils on Surface and Ground Water Systems*, Project Completion Report, Maine Water Resources Research Center, Univ. of Maine, Orono, 38 pp, Sept. 1972.
(NTIS: PB-213 173)

238. Mink, J. F., "Excessive Irrigation and the Soils and Ground Water of Oahu, Hawaii," *Science*, Vol. 135, No. 3504, pp 672-673, Feb, 23, 1962.

239. Tenorio, P. A., et al, *Identification of Return Irrigation Water in the Subsurface: Water Quality*, Technical Report No. 33,

Hawaii Water Resources Center, Univ. of Hawaii, Honolulu, 90 pp, Oct. 1969.
(NTIS: PB-189 171)

240. Tenorio, P. A., et al, *Identification of Irrigation Return Water in the Subsurface, Phase III: Kahuku, Oahu and Kahului and Lahaina, Maui,* Technical Report 44, Water Resources Research Center, Univ. of Hawaii, Honolulu, 53 pp, Dec. 1970.
(NTIS: PB-200 166)

241. Leonard, R. B., "Effect of Irrigation on the Chemical Quality of Ground and Surface Water, Cedar Bluff Irrigation District, Kansas," *Relationship of Agriculture to Soil and Water Pollution,* Cornell Univ. Conf. on Agricultural Waste Management, pp 147-163, 1970.

242. Law, J. P., Jr., et al, *Degradation of Water Quality in Irrigation Return Flows,* Bulletin B-684, R. S. Kerr Water Research Center, Oklahoma Agricultural Experiment Station, Ada; and Oklahoma State Univ., Stillwater, Dept. of Agronomy, 26 pp, Oct. 1970.

243. Alfaro, J. F., and D. W. Wilkins, *Model Study to Predict Salt Distribution and Concentration of Water in Soil Profiles,* Project Completion Report, Dept. of Agricultural Engineering, New Mexico State Univ., University Park, 16 pp, Dec. 1970.
(NTIS: PB-202 631)

244. Thomas, J. L., et al, *Quantity and Chemical Quality of Return Flow,* Report PRWG 77-1, Utah Water Research Lab., Logan, 94 pp, June 1971.
(NTIS: PB-201 004)

245. Law, J. P., Jr., *National Irrigation Return Flow Research and Development Program,* U.S. Environmental Protection Agency Report EPA-13030-GJS-12/71, Water Pollution Control Research Series, 30 pp, Dec. 1971.
(NTIS: PB-209 857)

246. Fitzsimmons, D. W., et al, "Nitrogen, Phosphorus, and Other Inorganic Materials in Waters in a Gravity-Irrigated Area," *Trans. Amer. Soc. of Agricultural Engineers,* Vol. 15, No. 2, pp 292-295, 1972.

247. Bonde, E. K., and P. Urone, "Plant Toxicants in Underground Water in Adams County, Colorado," *Soil Science,* Vol. 93, No. 5, pp 353-356, 1962.

248. McCarty, P. L., and P. H. King, "The Movement of Pesticides in Soils," *Industrial Waste Conf. Proceedings*, Vol. 21, Purdue Univ. Engineering Extension Series No. 121, pp 156-171, 1966.

249. Huggenberger, F., et al, "Adsorption and Mobility of Pesticides in Soil," *California Agriculture*, Vol. 27, No. 2, pp 8-10, 1973.

250. Eye, J. D., "Aqueous Transport of Dieldrin Residues in Soils," *Jour. Water Pollution Control Fed.*, Vol. 40, No. 8, pp R316-R332, 1968.

251. Boucher, R. R., and G. F. Lee, "Adsorption of Lindane and Dieldrin Pesticides on Unconsolidated Aquifer Sands," *Environmental Science and Technology*, Vol. 6, No. 6, pp 538-543, 1972.

252. Robertson, J. B., and I. Kahn, "The Infiltration of Aldrin Through Ottawa Sand Columns," *Geological Survey Research 1969*, U. S. Geological Survey Prof. Paper 650-C, pp C219-C223, 1969.

253. Dregne, H. E., et al, *Movement of 2,4-D in Soils*, Western Regional Research Project Progress Report, New Mexico Agricultural Experiment Station, University Park, 35 pp, Nov. 1969.

254. Mansell, R. S., and L. C. Hammond, *Movement and Adsorption of Pesticides in Sterilized Soil Columns*, Publication No. 15, Water Resources Research Center, Univ. of Florida, Gainesville, 63 pp, Aug. 1971.
(NTIS: PB-204 644)

255. Johnston, W. R., et al, "Insecticides in Tile Drainage Effluent," *Water Resources Research*, Vol. 3, No. 2, pp 525-537, 1967.

256. Working Group on Pesticides, *Ground Disposal of Pesticides: The Problem and Criteria for Guidelines*, Report WGP-DR-1, Rockville, Maryland, 55 pp, March 1970.
(NTIS: PB-197 144)

257. Schneider, A. D., et al, "Movement and Recovery of Herbicides in the Ogallala Aquifer," *The Ogallala Aquifer — A Symposium*, International Center for Arid and Semi-Arid Land Studies Special Report No. 39, Texas Tech Univ., Lubbock, pp 219-226, 1970.

258. Olsen, R. C., "Pesticides and Ground Water," *Water and Oregon's Ecology*, Fall Quarter 1970 Seminar Report WR 013.70, Oregon Water Resources Research Institute, Oregon State Univ., Corvallis, pp 79-81, Jan. 1971.

259. Swoboda, A. R., et al, "Distribution of DDT and Toxaphene in Houston Black Clay on Three Watersheds," *Environmental Science and Technology*, Vol. 5, pp 141-145, 1971.

260. Lewallen, M. J., "Pesticide Contamination of a Shallow Bored Well in the Southeastern Coastal Plains," *Ground Water*, Vol. 9, No. 6, pp 45-48, 1971.

261. Dixon, J. B., *Evaluation of Earthy Materials for Use in Decontamination of Water*, Water Resources Research Inst. Bull. 3, Auburn Univ., Auburn, Alabama, 74 pp, 1968.

262. Telfair, J. S., Jr., *The Pollution of Artesian Ground Waters in Suwannee and Orange Counties, Florida, by Artificial Recharge Through Drainage Wells*, Bureau of Sanitary Engineering Interim Report of Investigation, Florida State Board of Health, Tallahassee, 40 pp, 1948.

263. Reck, C. W., and E. J. Simmons, *Water Resources of the Buffalo-Niagara Falls Region*, U.S. Geological Survey Circular 173, 26 pp, 1952.

264. Sceva, J. E., *Liquid Waste Disposal in the Lava Terrane of Central Oregon*, Report No. FR-4, U.S. Federal Water Pollution Control Admin., Pacific Northwest Water Lab., Corvallis, Oregon, 66 pp, May 1968.
(NTIS: PB-191 874)

265. Abegglen, D. E., et al, *The Effects of Drain Wells on the Ground-Water Quality of the Snake River Plain*, Pamphlet 148, Idaho Bureau of Mines and Geology, Moscow, 51 pp, Oct. 1970.

266. Warner, D. L., *Deep-Well Injection of Liquid Waste — A Review of Existing Knowledge and an Evaluation of Research Needs*, U.S. Public Health Service Pub. 999-WP-21, 55 pp, 1965.

267. Eddy, G. E., "Subsurface Disposal of Industrial Wastes," *Interstate Oil Compact Commission Bulletin*, Vol. 9, No. 2, pp 71-79, 1967.

268. Research Committee, *Subsurface Disposal of Industrial Wastes,* Interstate Oil Compact Commission, 109 pp, June 1968.

269. Walker, W. R., and R. C. Stewart, "Deep-Well Disposal of Wastes," *Jour. Sanitary Engineering Div.,* Amer. Soc. of Civil Engineers, Vol. 94, No. SA 5, Paper 6171, pp 945-968, 1968.

270. Talbot, J. S., "Some Basic Factors in the Consideration and Installation of Deep Well Disposal Systems," *Water and Sewage Works,* Reference No. 1968, pp R213-R219, Nov. 1968.

271. Manning, J. C., "Deep Well Injection of Industrial Wastes," *Proceedings of 23rd Industrial Waste Conf.,* Part 2, Purdue Univ. Engineering Bulletin, Vol. 53, No. 2, pp 655-666, 1969.

272. Caswell, C. A., "Underground Waste Disposal," *Environmental Science and Technology,* Vol. 4, No. 8, pp 642-647, 1970.

273. Cook, T. D., Ed., *Underground Waste Management and Environmental Implications,* Memoir 18, Amer. Assoc. of Petroleum Geologists, Tulsa, Oklahoma, 412 pp, 1972.

274. Boegly, W. J., Jr., et al, *The Feasibility of Deep-Well Injection of Waste Brine from Inland Desalting Plants,* Research and Development Progress Report No. 432, Office of Saline Water, 76 pp, March 1969.
(NTIS: PB-203 852)

275. Rinne, W. W., "Need for Saline Groundwater Data to Advance Desalting Technology," *Water Resources Research,* Vol. 6, No. 5, pp 1482-1486, 1970.

276. Warner, D. L., *Deep Wells for Industrial Waste Injection in the United States — Summary of Data,* Water Pollution Control Research Series Publication No. WP-20-10, Federal Water Pollution Control Admin., 45 pp, 1967.

277. Warner, D. L., "Subsurface Disposal of Liquid Industrial Wastes by Deep-Well Injection," *Subsurface Disposal in Geologic Basins — A Study of Reservoir Strata,* Amer. Assoc. of Petroleum Geologists Memoir 10, pp 11-20, August 1968.

278. Anon., "Deep Injection Wells," *Water Well Journal,* Vol. 22, No. 8, pp 12-13, 1968.

279. Anon., "Deep Well Injection is Effective for Waste Disposal," *Environmental Science and Technology,* Vol. 2, No. 6, pp 406-410, 1968.

280. Selm, R. P., and B. T. Hulse, "Deep-Well Disposal of Industrial Wastes," *Proceedings 14th Industrial Wastes Conf.,* Purdue Univ. Engineering Extension Series No. 104, Purdue Univ. Engineering Bulletin, Vol. 44, No. 5, pp 566-586, 1959.

281. Stewart, R. S., "Techniques of Deep Well Disposal — A Safe and Efficient Method of Pollution Control," *Proceedings of the 15th Ontario Industrial Waste Conference,* pp 37-43, June 1968.

282. Warner, D. L., "Deep-Well Waste Injection Reaction with Aquifer Water," *Jour. Sanitary Engineering Div.,* Amer. Soc. of Civil Engineers, Vol. 92, No. SA 4, pp 45-69, 1966.

283. Marsh, J. H., "Design of Waste Disposal Wells," *Ground Water,* Vol. 6, No. 2, pp 4-8, 1968.

284. Slagle, K. A., and J. M. Stogner, "Oil Fields Yield New Deep-Well Disposal Technique," *Water and Sewage Works,* Vol. 116, No. 6, pp 238-244, 1969.

285. Rima, D. R., "Some Factors to be Considered in the Design of Waste Disposal Wells," *Proceedings of 8th Annual Sanitary and Water Resources Engineering Conference,* Technical Report No. 20, Dept. of Environment and Water Resources Engineering, Vanderbilt Univ., Nashville, Tenn., pp 119-127, 1969.

286. McLean, D. D., "Subsurface Disposal: Precautionary Measures," *Industrial Water Engineering,* Vol. 6, No. 8, pp 20-22, 1969.

287. Sheldrick, M. G., "Deep Well Disposal: Are Safeguards Being Ignored?" *Chemical Engineering,* Vol. 76, No. 7, pp 74-76, 78, 1969.

288. Piper, A. M., *Disposal of Liquid Wastes by Injection Underground — Neither Myth nor Millenium,* U.S. Geological Survey Circular 631, 15 pp, 1969.

289. National Industrial Pollution Control Council, *Waste Disposal in Deep Wells,* Sub-Council Report, Washington, D. C., 20 pp, Feb. 1971.

290. Miller, S. S., "Injection Wells Pose a Potential Threat," *Environmental Science and Technology,* Vol. 6, No. 2, pp 120-122, 1972.

291. Tofflemire, T. J., and G. P. Brezner, "Deep Well Injection of Waste Water," *Jour. Water Pollution Control Fed.,* Vol. 43, No. 7, pp 1468-1479, 1971.

292. Rudd, N., *Subsurface Liquid Waste Disposal and Its Feasibility in Pennsylvania,* Environmental Geol. Rept. 3, Penn. Geological Survey, 103 pp, 1972.

293. Otton, E. G., *Geologic and Hydrologic Factors Bearing on Subsurface Storage of Liquid Wastes in Maryland,* Maryland Geological Survey Report of Investigations No. 14, Johns Hopkins Univ., Baltimore, 39 pp, 1970.

294. Anon., "How to Bury a Major Pollution Problem," *Water Well Jour.,* Vol. 22, No. 8, p 20, 1968.

295. Cleary, E. J., and D. L. Warner, *Perspective on the Regulation of Underground Injection of Wastewaters,* Ohio River Valley Water Sanitation Commission, Cincinnati, 88 pp, Dec. 1969.

296. Cleary, E. J., and D. L. Warner, "Some Considerations in Underground Waste Water Disposal," *Jour. Amer. Water Works Assoc.,* Vol. 62, No. 8, pp 489-498, 1970.

297. Hundley, C. L., and J. T. Matulis, "Deep Well Disposal," *Ground Water,* Vol. 1, No. 2, pp 15-17, 33, 1963.

298. Hartman, C. D., "Deep Well Disposal of Steel Mill Wastes," *Jour. Water Pollution Control Fed.,* Vol. 40, No. 1, pp 95-100, 1968.

299. Bergstrom, R. E., *Feasibility of Subsurface Disposal of Industrial Wastes of Illinois,* Illinois State Geological Survey Circular 426, Urbana, 18 pp, 1968.

300. Bergstrom, R. E., "Feasibility Criteria for Subsurface Waste Disposal in Illinois," *Ground Water,* Vol. 6, No. 5, pp 5-9, 1968.

301. Smith, H. F., "Subsurface Storage and Disposal in Illinois," *Ground Water,* Vol. 9, No. 6, pp 20-28, 1971.

302. Berk, R. G., *Disposal Well Problems in Chicago and Bakersfield Areas,* Meeting Preprint 1302, National Water Resources Engineering Meeting, Phoenix, Amer. Soc. of Civil Engineers, 28 pp, Jan. 1971.

303. Jones, O. S., "Subsurface Disposal of Inland Oil Field Brines Conserves Fresh Water Supply," *Civil Engineering,* Vol. 17, No. 2, pp 60-63, 1947.

304. Grubbs, D. M., et al, *Conservation of Fresh-Water Resources by Deep-Well Disposal of Liquid Wastes,* Alabama Geological Survey and Univ. Alabama Natural Resources Center Report, 85 pp, May 1970.
(NTIS: PB-194 112)

305. Alverson, R. M., *Deep Well Disposal Study for Baldwin, Escambia and Mobile Counties, Alabama,* Alabama Geological Survey Circular 58, 49 pp, 1970.
(NTIS: PB-194 336)

306. Tucker, W. E., "Subsurface Disposal of Liquid Industrial Wastes in Alabama — A Current Status Report," *Ground Water,* Vol 9, No. 6, pp 10-19, 1971.

307. Batz, M. E., "Deep Well Disposal of Nylon Waste Water," *Chemical Engineering Progress,* Vol. 60, No. 10, pp 85-88, 1964.

308. Dean, B. T., "The Design and Operation of a Deep-Well Disposal System," *Jour. Water Pollution Control Fed.,* Vol. 37, No. 2, pp 245-254, 1965.

309. Barraclough, J. T., "Waste Injection Into a Deep Limestone in Northwestern Florida," *Ground Water,* Vol. 4, No. 1, pp 22-24, 1966.

310. Goolsby, D. A., "Hydrogeochemical Effects of Injecting Wastes Into a Limestone Aquifer Near Pensacola, Florida," *Ground Water,* Vol. 9, No. 1, pp 13-19, 1971.

311. Garcia-Bengochea, J. I., and R. O. Vernon, "Deep Well Disposal of Waste Waters in Saline Aquifers of South Florida," *Water Resources Research,* Vol. 6, No. 5, pp 1464-1470, 1970.

312. Vernon, R. O., *The Beneficial Uses of High Transmissivities in the Florida Subsurface for Water Storage and Waste Disposal,* Florida Bureau of Geology Information Circular No. 70, 39 pp, 1970.

313. Kaufman, M. I., "Subsurface Waste Injection, Florida," *Jour. Irrigation and Drainage Div.,* Amer. Soc. of Civil Engineers, Vol. 99, No. IR 1, pp 53-70, 1973.

314. de Ropp, H. W., "Chemical Waste Disposal at Victoria, Texas, Plant of DuPont Co.," *Sewage and Industrial Wastes,* Vol. 23, No. 2, pp 194-197, 1951.

315. Henkel, H. O., "Surface and Underground Disposal of Chemical Wastes at Victoria, Texas," *Sewage and Industrial Wastes,* Vol. 25, No. 9, pp 1044-1049, 1953.

316. Veir, B. B., "Celanese Deep-Well Disposal Practices," *Water and Sewage Works,* Vol. 116, No. 5, pp I/W 21-I/W 24, 1969.

317. Lockett, D. E., "Subsurface Disposal of Industrial Waste Water," *7th Industrial Water and Waste Conf. Proceedings,* Texas Water Pollution Control Assoc., Univ. of Texas, Austin, pp III-48-III-57, 1967.

318. McMillion, L. G., and B. W. Maxwell, *Determination of Pollutional Potential of the Ogallala Aquifer by Salt Water Injection,* R. S. Kerr Water Research Center, Ada, Oklahoma, 80 pp, June 1970. (NTIS: PB-202 227)

319. Evans, D. M., "The Denver Area Earthquakes and the Rocky Mountain Arsenal Disposal Well," *Mountain Geologist,* Vol. 3, No. 1, pp 23-26, 1966.

320. Evans, D. M., and A. Bradford, "Under the Rug," *Environment,* Vol. 11, No. 8, pp 3-13, 31, 1969.

321. McLean, D. D., *Subsurface Disposal of Liquid Wastes in Ontario,* Dept. of Energy and Resources Management, Petroleum Resources Section, Paper 68-2, 91 pp, Dec. 1968.

322. van Everdingen, R. O., and R. A. Freeze, *Subsurface Disposal of Waste in Canada,* Tech. Bull. No. 49, Inland Waters Branch, Dept. of the Environment, Ottawa, Canada, 64 pp, 1971.

323. University of California Sanitary Engineering Research Lab., *Investigation of Travel of Pollution,* California Water Pollution Control Board, Sacramento, Publication No. 11, 218 pp, 1954.

324. Krone, R. B., et al., "Direct Recharge of Ground Water with Sewage Effluents," *Jour. Sanitary Engineering Div.,* Amer. Soc. of Civil Engineers, Vol. 83, No. SA4, Paper 1335, 25 pp, 1957.

325. Mitchell, J. K., and W. R. Samples, *Reclamation of Waste Water for Well Injection,* Los Angeles County Flood Control District, Los Angeles, Calif., 250 pp, Feb. 1967.

326. Wesner, G. M., and D. C. Baier, "Injection of Reclaimed Waste-water into Confined Aquifers," *Jour. Amer. Water Works Assoc.,* Vol. 62, No. 3, pp 203-210, 1970.

327. Baier, D. C., and G. M. Wesner, "Reclaimed Waste Water for Groundwater Recharge," *Water Resources Bulletin,* Vol. 7, No. 5, pp 991-1001, 1971.

328. Cohen, P., and C. N. Durfor, "Design and Construction of a Unique Injection Well on Long Island, New York," *Geological Survey Research 1966,* U.S. Geological Survey Prof. Paper 550-D, pp D253-D257, 1966.

329. Baffa, J. J., and N. J. Bartilucci, "Wastewater Reclamation by Groundwater Recharge on Long Island," *Jour. Water Pollution Control Fed.,* Vol. 39, No. 3, Part 1, pp 431-445, 1967.

330. Cohen, P., and C. N. Durfor, "Artificial-Recharge Experiments Utilizing Renovated Sewage-Plant Effluent — A Feasibility Study at Bay Park, New York, U.S.A.," *International Assoc. of Scientific Hydrology,* Publication No. 72, pp 193-199, 1967.

331. Anon., "Replenishing the Aquifer with Treated Sewage Effluent," *Ground Water Age,* Vol. 2, No. 8, pp 30-35, 1968.

332. Peters, J. A., and J. L. Rose, "Water Conservation by Reclamation and Recharge," *Jour. Sanitary Engineering Div.,* Amer. Soc. of Civil Engineers, Vol. 94, No. SA4, pp 625-638, 1968.

333. Rose, J. L., "Injection of Treated Waste Water into Aquifers," *Water and Wastes Engineering,* Vol. 5, No. 10, pp 40-43, 1968.

334. Rose, J. L., "Advanced Waste Treatment in Nassau County, New York, Water Provided for Injection into Groundwater Aquifers," *Water and Wastes Engineering,* Vol. 7, No. 2, pp 38-39, 1970.

335. Vecchioli, J., "A Note on Bacterial Growth Around a Recharge Well at Bay Park, Long Island, New York," *Water Resources Research,* Vol. 6, No. 5, pp 1415-1419, 1970.

336. Vecchioli, J., "Experimental Injection of Tertiary-Treated Sewage in a Deep Well at Bay Park, Long Island, N.Y. — A Summary of Early Results," *New England Water Works Assoc. Jour.,* Vol. 86, No. 2, pp 87-103, 1972.

337. Vecchioli, J., et al, "Travel of Pollution-Indicator Bacteria Through the Magothy Aquifer, Long Island, New York," *Geological Survey Research 1972,* U.S. Geological Survey Prof. Paper 800-B, pp B237-B239, 1972.

338. Ehrlich, G. G., et al, "Microbiological Aspects of Ground-Water Recharge — Injection of Purified Chlorinated Sewage Effluent," *Geological Survey Research 1972,* U.S. Geological Survey Prof. Paper 800-B, pp B241-B245, 1972.

339. Williams, C. C., "Contamination of Deep Water Wells in Southeastern Kansas," *Kansas State Geological Survey Bulletin No. 76,* 1948 Reports of Studies, Part 2, pp 13-28, 1948.

340. Anon., "Ground Water Pollution is Costly," *Johnson Natl. Drillers Jour.,* Vol. 28, No. 4, pp 4-5, 1956.

341. U.S. Public Health Service, *Manual on Individual Water Supply Systems,* Public Health Service Publication No. 24, Washington, D.C., 121 pp, 1963.

342. Anon., "Bacteria Cause Many Problems," *Johnson Drillers Jour.,* Vol. 39, No. 4, pp 1-3, 1967.

343. Jorgensen, D. G., "An Aquifer Test Used to Investigate a Quality of Water Anomaly," *Ground Water,* Vol. 6, No. 6, pp 18-20, 1968.

344. Ham, H. H., "Water Wells and Ground-Water Contamination," *Bulletin of Assoc. of Engineering Geologists,* Vol. 8, No. 1, pp 79-90, 1971.

345. Jones, E. E., Jr., "Where Does Water Quality Improvement Begin?" *Ground Water*, Vol. 9, No. 3, pp 24-28, 1971.

346. Parker, G. G., "The Encroachment of Salt Water into Fresh," *Water, The Yearbook of Agriculture, 1955*, U.S. Dept. of Agriculture, pp 615-635, 1955.

347. Todd, D. K., "Salt Water Intrusion of Coastal Aquifers in the United States," *Intl. Assoc. of Scientific Hydrology*, Publication No. 52, pp 452-461, 1960.

348. Louisiana Water Resources Research Inst., *Salt-Water Encroachment into Aquifers*, Bulletin 3, Louisiana State Univ., Baton Rouge, 192 pp, Oct. 1968.

349. Task Committee on Saltwater Intrusion, "Salt-Water Intrusion in the United States," *Jour. Hydraulics Div.*, Amer. Soc. of Civil Engineers, Vol. 95, No. HY5, pp 1651-1669, 1969.

350. Lusczynski, N. J., and W. V. Swarzenski, *Salt-Water Encroachment in Southern Nassau and Southeastern Queens Counties, Long Island, New York*, U.S. Geological Survey Water-Supply Paper 1613-F, 76 pp, 1966.

351. Cohen, P., and G. E. Kimmel, "Status of Salt-Water Encroachment in 1969 and in Southern Nassau and Southeastern Queens Counties, Long Island, New York," *Geological Survey Research 1970*, U.S. Geological Survey Prof. Paper 700-D, pp D281-D286, 1970.

352. Woodruff, K. D., *The Occurrence of Saline Ground Water in Delaware Aquifers*, Delaware Geological Survey Report of Investigations No. 13, 45 pp, 1969.

353. McCollum, M. J., "Salt-Water Movement in the Principal Artesian Aquifer of the Savannah Area, Georgia and South Carolina," *Ground Water*, Vol. 2, No. 4, pp 4-8, 1964.

354. Counts, H. B., and E. Donsky, *Salt-Water Encroachment, Geology, and Ground-Water Resources of Savannah Area, Georgia and South Carolina*, U.S. Geological Survey Water-Supply Paper 1611, 100 pp, 1964.

355. McCollum, M. J., and H. B. Counts, *Relation of Salt-Water Encroachment to the Major Aquifer Zones, Savannah Area, Georgia and South Carolina,* U.S. Geological Survey Water-Supply Paper 1613-D, 26 pp, 1964.

356. Siple, G. E., "Salt-Water Encroachment of Tertiary Limestone Along Coastal South Carolina," *Hydrology of Fractured Rocks,* Vol. 2, Proc. of Dubrovnik Symposium, Intl. Assoc. of Scientific Hydrology, Publication 74, pp 439-453, 1967.

357. Back, W., et al, "Carbon-14 Ages Related to Occurrence of Salt Water," *Jour. Hydraulics Div.,* Amer. Soc. of Civil Engineers, Vol. 96, No. HY11, Paper 7702, pp 2325-2336, 1970.

358. Wait, R. L., "Notes on the Position of a Phosphate Zone and Its Relation to Groundwater in Coastal Georgia," *Geological Survey Research 1970,* U.S. Geological Survey Prof. Paper 700-C, pp C202-C205, 1970.

359. Gregg, D. O., "Protective Pumping to Reduce Aquifer Pollution, Glynn County, Georgia," *Ground Water,* Vol. 9, No. 5, pp 21-29, 1971.

360. Wait, R. L., and J. T. Callahan, "Relations of Fresh and Salty Ground Water Along the Southeastern U.S. Atlantic Coast," *Ground Water,* Vol. 3, No. 4, pp 3-17, 1965.

361. Black, A. P., et al, *Salt Water Intrusion in Florida – 1953,* Florida Div. of Water Survey and Research Paper No. 9, 38 pp, 1953.

362. Klein, H., *Interim Report on Salt-Water Encroachment in Dade County, Florida,* Florida Geological Survey Information Circular No. 9, 17 pp, 1957.

363. Kohout, F. A., "Case History of Salt Water Encroachment Caused by a Storm Sewer in Miami," *Jour. Amer. Water Works Assoc.,* Vol. 53, No. 11, pp 1406-1416, 1961.

364. Kohout, F. A., and H. Klein, "Effect of Pulse Recharge on the Zone of Diffusion in the Biscayne Aquifer," *Artificial Recharge and Management of Aquifers,* Intl. Assoc. of Scientific Hydrology, Publication No. 72, pp 252-270, 1967.

365. Stringfield, V. T., and H. E. LeGrand, "Relation of Sea Water to Fresh Water in Carbonate Rocks in Coastal Areas, with Special Reference to Florida, U.S.A., and Cephalonia (Kephallinia), Greece," *Jour. of Hydrology,* Vol. 9, No. 4, pp 387-404, 1969.

366. Sproul, C. R., et al, *Saline-Water Intrusion from Deep Artesian Sources in the McGregor Isles Area of Lee County, Florida,* Florida, Dept. of Natural Resources, Div. of Interior Resources, Bureau of Geology Information Circular No. 75, Tallahassee, 30 pp, 1972.

367. Anon., "Underground Storage of Storm Water Runoff," *Ground Water Age,* Vol. 7, No. 8, pp 29, 32, 36-37, 1973.

368. Meyer, R. R., and J. R. Rollo, *Salt Water Encroachment, Baton Rouge Area, Louisiana,* Louisiana Geological Survey Water Resources Pamphlet No. 17, 9 pp, 1965.

369. Rollo, J. R., *Salt Water Encroachment in Aquifers of the Baton Rouge Area, Louisiana,* Louisiana Dept. of Conservation, Geological Survey, and Dept. of Public Works, Water Resources Bulletin No. 13, 45 pp, August 1969.

370. Smith, C. G., Jr., *Geohydrology of the Shallow Aquifers of Baton Rouge, Louisiana,* Louisiana Water Resources Research Institute Bulletin GT-4, Louisiana State Univ., Baton Rouge, 31 pp, Oct. 1969.

371. Long, R. A., *Feasibility of a Scavenger-Well System as a Solution to the Problem of Vertical Salt-Water Encroachment,* Louisiana Geological Survey Water Resources Pamphlet 15, 27 pp, 1965.

372. Harder, A. H., et al, *Effects of Ground-Water Withdrawals on Water Levels and Salt-Water Encroachment in Southwestern Louisiana,* Louisiana Geological Survey Water Resources Bulletin No. 10, 56 pp, 1967.

373. Winslow, A. G., and W. W. Doyel, *Salt Water and Its Relation to Fresh Ground Water in Harris County, Texas,* Texas Board of Water Engineers Bulletin 5409, 37 pp, 1954.

374. Winslow, A. G., et al, *Salt Water and Its Relation to Fresh Ground Water in Harris County, Texas,* U.S. Geological Survey Water-Supply Paper 1360-F, 33 pp, 1957.

375. Piper, A. M., et al, *Native and Contaminated Ground Waters in the Long Beach-Santa Ana Area, California,* U.S. Geological Survey Water-Supply Paper 1136, 320 pp, 1953.

376. California Dept. of Water Resources, *Sea Water Intrusion in California,* Bulletin 63, 91 pp, 1958.

377. California Dept. of Water Resources, *Sea Water Intrusion in California,* Bulletin No. 63, Appendix B, Los Angeles County Flood Control District Report, 76 pp, 1957.

378. California Dept. of Water Resources, *Sea Water Intrusion in California,* Bulletin No. 63, Appendices C, D, and E, 245 pp, 1960.

379. California Dept. of Water Resources, *San Dieguito River Investigation,* Bulletin 72, Vol. 1, 197 pp, 1959.

380. California Dept. of Water Resources, *Intrusion of Salt Water into Ground Water Basins of Southern Alameda County,* Bulletin 81, 44 pp, 1960.

381. California Dept. of Water Resources, *Sea-Water Intrusion, Oxnard Plain of Ventura County,* Bulletin 63-1, 59 pp, 1965.

382. California Dept. of Water Resources, *Ground Water Basin Protection Projects: Oxnard Basin Experimental Extraction-Type Barrier,* Bulletin 146-7, 157 pp, 1970.

383. California Dept. of Water Resources, *Sea-Water Intrusion: Aquitards in the Coastal Ground Water Basin of Oxnard Plain, Ventura County,* Bulletin 63-4, 569 pp, 1971.

384. California Dept. of Water Resources, *Ground Water Basin Protection Projects: Santa Ana Gap Salinity Barrier, Orange County,* Bulletin 147-1, 178 pp, Dec. 1966.

385. California Dept. of Water Resources, *Sea-Water Intrusion: Bolsa-Sunset Area, Orange County,* Bulletin No. 63-2, 167 pp, Jan. 1968.

386. California Dept. of Water Resources, *Sea Water Intrusion: Pismo-Guadalupe Area,* Bulletin 63-3, 76 pp, 1970.

387. California Dept. of Water Resources, *Sea Water Intrusion: Morro Bay Area, San Luis Obispo County,* Bulletin 63-6, 104 pp, 1972.

388. Brown, P. G., "Potential Uses of Reclaimed Municipal Waste Water," *Proceedings of 2d Biennial Microbiology Symposium*, Amer. Petroleum Inst., Paper 18, pp 144-155, 1965.

389. Bruington, A. E., "Control of Sea-Water Intrusion in a Ground-Water Aquifer," *Ground Water*, Vol. 7, No. 3, pp 9-14, 1969.

390. McIlwain, R. R., et al, *West Coast Basin Barrier Project 1967– 1969*, Los Angeles County Flood Control District, Los Angeles, Calif., 30 pp, June 1970.

391. Alves, E., Jr., and D. B. Hunt, *Alamitos Barrier Project Report for 1968-69*, Report to Los Angeles County Flood Control District, 30 pp, Oct. 1969.

392. Moore, C. V., and J. H. Snyder, "Some Legal and Economic Implications of Sea Water Intrusion – A Case Study of Ground Water Management," *Natural Resources Jour.*, Vol. 9, No. 3, pp 401-419, 1969.

393. Lau, L. S., *Dynamic and Static Studies of Seawater Intrusion*, Technical Report No. 3, Water Resources Research Center, Univ. of Hawaii, Honolulu, 31 pp, Feb. 1967.

394. Todd, D. K., and C. F. Meyer, "Hydrology and Geology of the Honolulu Aquifer," *Jour. Hydraulics Div.*, Amer. Soc. of Civil Engineers, Vol. 97, No. HY 2, pp 233-256, 1971.

395. Tremblay, J. J., et al, "Salt Water Intrusion in the Summerside Area, P.E.I.," *Ground Water*, Vol. 11, No. 2, pp 21-27, 1973.

396. Cooper, H. H., Jr., et al, *Sea Water in Coastal Aquifers*, U.S. Geological Survey Water-Supply Paper 1613-C, 84 pp, 1964.

397. Carstens, M. R., and G. D. May, *Salt-Water Intrusion Effect of a Fresh-Water Canal*, Water Resources Center, Georgia Inst. of Technology, Atlanta, 49 pp, May 1967.

398. Charmonman, S., et al, "A Fresh-Water Canal as a Barrier to Salt-Water Intrusion," *Artificial Recharge and Management of Aquifers*, Intl. Assoc. of Scientific Hydrology, Publication No. 72, pp 374-382, 1967.

399. Carlson, E. J., and P. F. Enger, "Removal of Saline Water from Aquifers," *Proceedings of 13th Congress of the International Assoc. for Hydraulic Research,* Vol. 4, (Subject D), Science Council of Japan, Kyoto, pp 121-134, 1969.

400. Pinder, G. F., and H. H. Cooper, "A Numerical Technique for Calculating the Transient Position of the Saltwater Front," *Water Resources Research,* Vol. 6, No. 3, pp 875-882, 1970.

401. Kashef, A. A. I., "Model Studies of Salt Water Intrusion," *Water Resources Bulletin,* Vol. 6, No. 6, pp 944-967, 1970.

402. Kashef, A. A. I., "On the Management of Groundwater in Coastal Aquifers," *Ground Water,* Vol. 9, No. 2, pp 12-20, 1971.

403. Kashef, A. A. I., "What Do We Know About Salt Water Intrusion?" *Water Resources Bulletin,* Vol. 8, No. 2, pp 282-293, 1972.

404. Peek, H. M., "Effects of Large-Scale Mining Withdrawals of Ground Water," *Ground Water,* Vol. 7, No. 4, pp 12-20, 1969.

405. Hennighausen, F. H., "Change of Chloride Content of Water in Response to Pumping in the Artesian Aquifer in the Roswell-East Grand Plains Area, Chaves County, New Mexico," *Saline Water,* Contribution 13, AAAS Comm. on Desert and Arid Zone Research, pp 71-86, 1970.

406. Norris, S. E., "Effects on Ground-Water Quality and Induced Infiltration of Wastes Disposed into the Hocking River at Lancaster, Ohio," *Ground Water,* Vol. 5, No. 3, pp 15-19, 1967.

407. Klaer, F. H., Jr., "Bacteriological and Chemical Factors in Induced Infiltration," *Ground Water,* Vol. 1, No. 1, pp 38-43, 1963.

408. Preul, H. C., and L. V. Pupal, "Effect of River Water Quality on an Adjacent Aquifer," *Systems Approach to Water Quality in the Great Lakes,* Proceedings of 3rd Annual Symposium on Water Resources Research, Ohio State Univ., pp 73-96, Sept. 1967.

409. Randall, A. D., "Movement of Bacteria from a River to a Municipal Well — A Case History," *Jour. Amer. Water Works Assoc.,* Vol. 62, No. 11, Part 1, pp 716-720, 1970.

410. American Water Works Assoc., "Control of Underground Waste Disposal," *Jour. Amer, Water Works Assoc.,* Vol. 44, No. 8, pp 685-689, 1952.

411. Amer. Water Works Assoc., "Underground Waste Disposal and Control," *Jour. Amer. Water Works Assoc.,* Vol. 49, No. 10, pp 1334-1342, 1957.

412. Amer. Water Works Assoc., "Underground Waste Disposal and Ground-Water Contamination," *Jour. Amer. Water Works Assoc.,* Vol. 52, No. 5, pp 619-622, 1960.

413. Task Group Report, "Survey of Ground Water Contamination and Waste Disposal Practices," *Jour. Amer. Water Works Assoc.,* Vol. 52, No. 9, pp 1211-1219, 1960.

414. World Health Organization, "Pollution of Ground Water (excerpt)," *Jour. Amer. Water Works Assoc.,* Vol. 49, No. 4, pp 392-396, 1957.

415. Rorabaugh, M. I., "Problems of Waste Disposal and Ground Water Quality," *Jour. Amer. Water Works Assoc.,* Vol. 52, No. 8, pp 979-982, 1960.

416. Bolton, P., "Prevention of Water Source Contamination," *Jour. Amer. Water Works Assoc.,* Vol. 53, No. 10, pp 1243-1250, 1961.

417. Meron, A., and H. F. Ludwig, "Salt Balances in Ground Water," *Jour. Sanitary Engineering Div.,* Amer. Soc. of Civil Engineers, Vol. 89, No. SA 3, pp 41-61, 1963.

418. LeGrand, H. E., "Management Aspects of Groundwater Contamination," *Jour. Water Pollution Control Fed.,* Vol. 36, No. 9, pp 1133-1145, 1964.

419. LeGrand, H. E., "Environmental Framework of Groundwater Contamination," *Ground Water,* Vol. 3, No. 2, pp 11-15, 1965.

420. LeGrand, H. E., "A Broad View of Waste Disposal in the Ground," *Water and Sewage Works,* Vol. 114, 1967 Reference Number, pp R167-R170, R179-R180, Nov. 1967.

421. LeGrand, H. E., "Role of Ground Water Contamination in Water Management," *Jour. Amer. Water Works Assoc.,* Vol. 59, No. 5, pp 557-565, 1967.

422. LeGrand, H. E., "Patterns of Contaminated Zones of Water in the Ground," *Water Resources Research*, Vol. 1, No. 1, pp 83-95, 1965.

423. Rainwater, F. H., "Natural Ground Water Quality Problems," *Jour. Soil and Water Conservation*, Vol. 20, No. 6, pp 254-255, 1965.

424. Oltman, R. E., "Research Under P.L. 88-379 in Quality Aspects of Water," *Ground Water*, Vol. 4, No. 2, pp 5-10, 1966.

425. McGauhey, P. H., "Manmade Contamination Hazards," *Ground Water*, Vol. 6, No. 3, pp 10-13, 1968.

426. Anon., "Pollution of Groundwater," *Legal Control of Water Pollution*, Univ. of California, Davis, Law Rev., pp 141-165, 1969.

427. Editors of Water Well Journal, "Groundwater Pollution – The Authoritative Primer," *Water Well Jour.*, Vol. 24, No. 7, pp 31-67, 1970.

428. MacKenthun, K. M., and L. E. Keup, "Biological Problems Encountered in Water Supplies," *Jour. Amer. Water Works Assoc.*, Vol. 62, No. 8, pp 520-526, 1970.

429. Kazmann, R. G., "Exotic Uses of Aquifers," *Jour. Irrigation and Drainage Div.*, Amer. Soc. of Civil Engineers, Vol. 97, No. IR 3, Paper 8352, pp 515-522, 1971.

430. Pettyjohn, W. A., "Good Coffee Water Needs Body," *Ground Water*, Vol. 10, No. 5, pp 47-49, 1972.

431. Callahan, J. T., "The Role of the U.S. Geological Survey in Waste Disposal Monitoring," *Ground Water*, Vol. 10, No. 3, pp 6-9, 1972.

432. Wood, L. A., "Groundwater Degradation – Causes and Cures," *Proceedings of the 14th Water Quality Conference on Groundwater Quality and Treatment*, Univ. of Illinois Dept. of Civil Engineering, Univ. of Illinois Bulletin, Vol. 69, No. 120, pp 19-25, May 1972.

433. Lewicke, C. K., "Ground Water Pollution and Conservation," *Environmental Science and Technology*, Vol. 6, No. 3, pp 213-215, 1972.

434. Hughes, G. M., and K. Cartwright, "Scientific and Administrative Criteria for Shallow Waste Disposal," *Civil Engineering*, Vol. 42, No. 3, pp 70-73, 1972.

435. Lin, S., *Nonpoint Rural Sources of Water Pollution*, Circular 111, Illinois State Water Survey, Urbana, 36 pp, 1972.

436. Walker, W. H., "Where Have All the Toxic Chemicals Gone?" *Ground Water*, Vol. 11, No. 2, pp 11-20, 1973.

437. Motts, W. S., "Groundwater Contamination and Hydrochemical Facies of Shallow Aquifers in Massachusetts," *Water Resources Research Center Symposium Proceedings*, Univ. of Massachusetts, Publication No. 3, pp 29-38, June 1967.

438. Motts, W. S., and M. Saines, *The Occurrence and Characteristics of Ground-Water Contamination in Massachusetts*, Water Resources Research Publication No. 7, Univ. of Massachusetts, Amherst, 70 pp, Jan. 1969.

439. Perlmutter, N. M., and A. A. Guerrera, *Detergents and Associated Contaminants in Ground Water at Three Public-Supply Well Fields in Southwestern Suffolk County, Long Island, New York*, U.S. Geological Survey Water Supply Paper 2001-B, 22 pp, 1970.

440. Chemerys, J. C., "Effect of Urban Development on Quality of Ground Water, Raleigh, N. C.," *Geological Survey Research 1967*, U.S. Geological Survey Prof. Paper 575-B, pp B212-B216, 1967.

441. Parizek, R. R., "Land Use Problems in Carbonate Terranes," *Hydrogeology and Geochemistry of Folded and Faulted Rocks of the Central Appalachian Type and Related Land Use Problems* (R. R. Parizek, et al, editors), Guidebook to Field Trips No. 11, Geological Soc. of America, Washington, D. C., pp. 135-180, Nov. 1971.

442. Wilmoth, B. M., "Salty Ground Water and Meteoric Flushing of Contaminated Aquifers in West Virginia," *Ground Water*, Vol. 10, No. 1, pp 99-106, 1972.

443. Billings, N., "Ground-Water Pollution in Michigan," *Sewage and Industrial Wastes*, Vol. 22, No. 12, pp 1596-1600, 1950.

444. Deutsch, M., *Ground Water Contamination and Legal Controls in Michigan,* U.S. Geological Survey Water-Supply Paper 1691, 79 pp, 1963.

445. Burt, E. M., "The Use, Abuse and Recovery of a Glacial Aquifer," *Ground Water,* Vol. 10, No. 1, pp 65-71, 1972.

446. Jordan, D. G., "Ground-Water Contamination in Indiana," *Jour. Amer. Water Works Assoc.,* Vol. 54, No. 10, pp 1213-1220, 1962.

447. George, A. I., "Pollution of Karst Aquifers," *Water Well Jour.,* Vol. 27, No. 8, pp 29-32, 1973.

448. Hackett, J. E., "Ground-Water Contamination in an Urban Environment," *Ground Water,* Vol. 3, No. 3, pp 27-30, 1965.

449. Walker, W. H., "Illinois Ground Water Pollution," *Jour. Amer. Water Works Assoc.,* Vol. 61, No. 1, pp 31-40, 1969.

450. Wikre, D., "Ground Water Pollution Problems in Minnesota," *Proceedings of Conf. Toward a Statewide Ground Water Quality Information System,* Water Resources Research Center, Univ. of Minnesota, Minneapolis, pp 59-78, Feb. 1973.

451. Williams, J. H., "Can Groundwater Pollution be Avoided?" *Ground Water,* Vol. 7, No. 2, pp 21-23, 1969.

452. Foley, F. C., and B. F. Latta, "Salt Water Encroachment from Industrial Operations," *Amer. Water Resources Assoc. Proceedings,* Series No. 2, pp 56-63, 1966.

453. Gahr, W. N., "Contamination of Ground Water — Vicinity of Denver," *Symposium on Water Improvement,* Amer. Assoc. for the Adv. Science, 128th Annual Meeting, Denver, Colorado, pp 9-20, 1961.

454. Page, H. G., and C. H. Wayman, "Removal of ABS and Other Sewage Components by Infiltration Through Soils," *Ground Water,* Vol. 4, No. 1, pp 10-17, 1966.

455. White, N. F., and D. K. Sunada, *Ground-Water Quality Study of Severance Basin. Weld County, Colorado,* Water Resources Research Report, Colorado State Univ., Fort Collins, 29 pp, April 1966.

456. U.S. Federal Water Pollution Control Admin., *Ground Water Pollution in the Middle and Lower South Platte River Basin of Colorado,* South Platte River Basin Project PR-9, 41 pp, 1968.

457. Handy, A. H., et al, "Changes in Chemical Quality of Ground Water in Three Areas in the Great Basin, Utah," *Geological Survey Research 1969,* U.S. Geological Survey Prof. Paper 650-D, pp D228-D234, 1969.

458. Brown, S. G., *Problems of Utilizing Ground Water in the West-Side Business District of Portland, Oregon,* U.S. Geological Survey Water-Supply Paper 1619-O, 42 pp, 1963.

459. California Dept. of Water Resources, *Water Quality and Water Quality Problems, Ventura County,* Bulletin 75, 2 Vols., 195 pp, 1959.

460. California Dept. of Water Resources, *Lower San Joaquin Valley Water Quality Investigation,* Bulletin 89, 189 pp, 1960.

461. California Dept. of Water Resources, *Ground Water Quality Studies in Mojave River Valley in the Vicinity of Barstow – San Bernardino County,* Report to the Lahontan Regional Water Pollution Control Board, No. 6, 60 pp, June 1960.

462. Evenson, R. E., *Suitability of Irrigation Water and Changes in Ground-Water Quality in the Lompoc Subarea of the Santa Ynez River Basin, Santa Barbara County, California,* U.S. Geological Survey Water-Supply Paper 1809-S, 20 pp, 1965.

463. California Dept. of Water Resources, *Ground Water Occurrence and Quality: San Diego Region,* Bulletin 106-2, 2 Vols., 235 pp, 1967.

464. California Dept. of Water Resources, *Santa Clara River Valley Water Quality Study,* Ground Water Basin Protection Projects Report, 128 pp, May 1968.

465. Orlob, G. T., and B. B. Dendy, "Systems Approach to Water Quality Management," *Jour. Hydraulics Div.,* Amer. Soc. of Civil Engineers, Vol. 99, No. HY 4, pp 573-587, 1973.

466. Ritter, C., and W. J. Hausler, Jr., "Yearly Variation in Sanitary Quality of Well Water," *Amer. Jour. of Public Health,* Vol. 51, No. 9, pp 1347-1357, 1961.

467. Drewry, W. A., and R. Eliassen, "Virus Movement in Groundwater," *Jour. Water Pollution Control Fed.*, Vol. 40, No. 8, Part 2, pp R257-R271, 1968.

468. Drewry, W. A., *Virus Movement in Groundwater Systems*, Report No. Pub-4, Water Resources Research Center, Univ. of Arkansas, Fayetteville, 85 pp, Sept. 1969. (NTIS: PB-188 285)

469. Carlson, G. F., et al, "Virus Inactivation on Clay Particles in Natural Waters," *Jour. Water Pollution Control Fed.*, Vol. 40, No. 2, pp R89-R106, 1968.

470. Tanimoto, R. M., et al, *Migration of Bacteriophage T4 in Percolating Water Through Selected Oahu Soils*, Technical Report No. 20, Water Resources Research Center, Univ. of Hawaii, Honolulu, 45 pp, June 1968.

471. Hori, D. H., et al, *Migration of Poliovirus Type 2 in Percolating Water Through Selected Oahu Soils*, Technical Report No. 36, Hawaii Water Resources Research Center, Univ. of Hawaii, Honolulu, 40 pp, Jan. 1970.

472. Bigbee, P. D., and R. G. Taylor, *Pollution Studies of the Regional Ogallala Aquifer at Portales, New Mexico*, Partial Completion Report 005, New Mexico Water Resources Research Inst., Las Cruces, 30 pp, Aug. 1972. (NTIS:  PB-212 264)

473. Vander Velde, T. L., "Poliovirus in a Water Supply," *Jour. Amer. Water Works Assoc.*, Vol. 65, No. 5, pp 345-346, May 1973.

474. Deluty, J., "Synthetic Detergents in Well Water," *Public Health Reports*, Vol. 75, No. 1, pp 75-77, 1960.

475. Perlmutter, N. M., et al, "Contamination of Ground Water by Detergents in a Suburban Environment — South Farmingdale Area, Long Island, New York," *Geological Survey Research 1964*, U.S. Geological Survey Prof. Paper 501-C, pp C170-C175, 1964.

476. Perlmutter, N. M., and E. Koch, "Preliminary Findings on the Detergent and Phosphate Contents of Water of Southern Nassau County, New York," *Geological Survey Research 1971*, U.S. Geological Survey Prof. Paper 750-D, pp D171-D177, 1971.

477. Wayman, C., et al, *Behavior of Surfactants and Other Detergent Components in Water and Soil-Water Environments,* U.S. Federal Housing Administration Technical Studies, Publication FHA 532, 136 pp, Feb. 1965.

478. Holloway, H. D., *Investigation of Ground-Water Contamination, Rhineland Area, Knox County, Texas,* Texas Water Comm. Bulletin 6521, 39 pp, 1965.

479. California Dept. of Water Resources, *Dispersion and Persistence of Synthetic Detergents in Ground Water, San Bernardino and Riverside Counties,* Water Quality Control Board Publication 30, 45 pp, 1965.

480. Dunlap, W. J., et al, *Investigations Concerning Probable Impact of Nitrilotriacetic Acid on Ground Water,* U.S. Environmental Protection Agency, Water Pollution Control Research Series, 51 pp, Nov. 1971.
(NTIS:    PB-208 433)

481. Dunlap, W. J., et al, "Probable Impact of NTA on Ground Water," *Ground Water,* Vol. 10, No. 1, pp 107-117, 1972.

482. Klein, S. A., *The Fate of NTA in Septic-Tank and Oxidation Pond Systems,* Sanitary Engineering Research Lab., SERL Report No. 71-4, Univ. of California, Berkeley, 96 pp, 1971.

483. Klein, S. A., *The Fate of Detergents in Septic Tank Systems and Oxidation Ponds,* Sanitary Engineering Research Lab., SERL Report No. 64-1, Univ. of California, Berkeley, 79 pp, 1964.

484. Klein, S. A., and D. Jenkins, *The Fate of Carboxymethyloxysuccinate in Septic Tank and Oxidation Pond Systems,* Sanitary Engineering Research Lab., SERL Report No. 72-10, Univ. of California, Berkeley, 55 pp, 1972.

485. Snoeyink, Y., and V. Griffin, *Nitrate and Water Supply: Source and Control,* College of Engineering Publication, Univ. of Illinois, 195 pp, 1970.

486. Sepp, E., *Nitrogen Cycle in Ground Water,* Bureau of Sanitary Engineering, California Dept. of Public Health, 23 pp, 1970.

487. Black, C. A., "Behavior of Soil and Fertilizer Phosphorus in Relation to Water Pollution," *Agricultural Practices and Water Quality,* Willrich, T. L., and G. E. Smith, Ed., Iowa State Univ. Press, Ames, pp 72-93, 1970.

488. Goldberg, M. C., "Sources of Nitrogen in Water Supplies," *Agricultural Practices and Water Quality,* Willrich and Smith (Eds.), Iowa State Univ. Press, Ames, pp 94-124, 1970.

489. Keeny, D. R., "Nitrates in Plants and Waters," *Jour. Milk and Food Technology,* Vol. 33, No. 10, pp 425-432, 1970.

490. Viets, F. J., and R. H. Hageman, *Factors Affecting the Accumulation of Nitrate in Soil, Water and Plants,* Agriculture Handbook 413, Agricultural Research Service, U.S. Dept. of Agriculture, Washington, D. C., 63 pp, 1971.

491. Goldberg, M. C., "Sources of Nitrogen in Water Supplies," *Role of Agriculture in Clean Water,* Iowa State Univ. Press, Ames, pp 94-124, 1971.

492. Winton, E. F., et al, "Nitrate in Drinking Water: Public Health Aspects," *Jour. Amer. Water Works Assoc.,* Vol. 63, No. 2, pp 95-98, 1971.

493. Lance, J. C., "Nitrogen Removal by Soil Mechanisms," *Jour. Water Pollution Control Fed.,* Vol. 44, No. 7, pp 1352-1361, 1972.

494. Kimmel, G. E., "Nitrogen Content of Ground Water in Kings County, Long Island, N. Y.," *Geological Survey Research 1972,* U.S. Geological Survey Prof. Paper 800-D, 1972.

495. Miller, J. C., *Nitrate Contamination of the Water-Table Aquifer in Delaware,* Report of Investigations No. 20, Delaware Geological Survey, Newark, 36 pp, May 1972.

496. Walker, E. H., *Ground-Water Resources of the Hopkinsville Quadrangle, Kentucky,* U.S. Geological Survey Water-Supply Paper 1328, 98 pp, 1956.

497. Peele, T. C., and J. T. Gillingham, *Influence of Fertilization and Crops on Nitrate Content of Groundwater and Tile Drainage Effluent,* Report No. 33, Water Resources Research Inst., Clemson Univ., Clemson, S. C., 19 pp, 1972.

498. Larson, T. E., and L. M. Henley, *Occurrence of Nitrate in Well Waters,* Research Report 1, Univ. of Illinois Water Resources Center, 8 pp, 1966.

499. Dawes, J. H., et al, "Nitrate Pollution of Water," *Frontiers in Conservation,* Proceedings of 24th Annual Meeting of the Soil Conservation Soc. of Amer., Colorado State Univ., Fort Collins, pp 94-102, 1970.

500. Walker, W. H., "Ground-Water Nitrate Pollution in Rural Areas," *Ground Water,* Vol. 11, No. 5, pp 19-22, 1973.

501. Murphy, S., and J. W. Gosch, *Nitrate Accumulation in Kansas Groundwater,* Project Completion Report, Kansas Water Resources Research Inst., Kansas State Univ., Manhattan, 56 pp, March 1970. (NTIS: PB-191 066)

502. Engberg, R. A., *The Nitrate Hazard in Well Water with Special Reference to Holt County, Nebraska,* Nebraska Water Survey Paper 21, Conservation and Survey Div., Univ. of Nebraska, Lincoln, 18 pp, 1967.

503. Witzel, S. A., et al, *Nitrogen Cycle in Surface and Subsurface Waters,* Technical Completion Report, Univ. of Wisconsin, Water Resources Center, 65 pp, Dec. 1968. (NTIS: PB-188 815)

504. Olsen, R. J., *Effect of Various Factors on Movement of Nitrate Nitrogen in Soil Profiles and on Transformations of Soil Nitrogen,* Water Resources Center Report 1969, Univ. of Wisconsin, 79 pp, 1969.

505. Crabtree, K. T., *Nitrate Variation in Groundwater,* Supplementary Report, Univ. of Wisconsin, Madison, Water Resources Center, 60 pp, 1970. (NTIS: PB-193 707)

506. Anon., "Fertilizers and Feedlots — What Role in Groundwater Pollution?" *Agricultural Research,* Vol. 18, No. 6, pp 14-15, 1969.

507. Taylor, R. G., and P. D. Bigbee, *Fluctuations in Nitrate Concentrations Utilized as an Assessment of Agricultural Contamination to an Aquifer of a Semiarid Climatic Region,* Partial Completion

Report 006, New Mexico Water Resources Research Institute,
Las Cruces, 12 pp, August 1972.
(NTIS:   PB-212 265)

508. Blanchar, R. W., and C. W. Kao, *Effects of Recent and Past Phosphate Fertilization on the Amount of Phosphorus Percolating Through Soil Profiles into Subsurface Waters*, Project Completion Report, Missouri Water Resources Research Center, Columbia, 106 pp, July 1971.
(NTIS:   PB-204 702)

509. Navone, R., et al, "Nitrogen Content of Ground Water in Southern California," *Jour. Amer. Water Works Assoc.*, Vol. 55, pp 615-618, 1963.

510. California Bureau of Sanitary Engineering, *Occurrence of Nitrate in Ground Water Supplies in Southern California*, California State Dept. of Public Health, 7 pp, Feb. 1963.

511. U.S. Federal Water Quality Admin., *Collected Papers Regarding Nitrates in Agricultural Waste Waters*, Federal Water Quality Admin. Water Pollution Control Research Series 13030 ELY 12/69, San Francisco, California, 186 pp, Dec. 1969.
(NTIS:   PB-197 595)

512. Shaffer, M. J., et al, "Predicting Changes in Nitrogenous Compounds in Soil-Water Systems," *Collected Papers Regarding Nitrates in Agricultural Waste Waters*, Federal Water Quality Admin. Water Pollution Control Research Series 13030 ELY 12/69, pp 15-28, Dec. 1969.

513. Ward, P. C., "Existing Levels of Nitrates in Water — The California Situation," *Proceedings of 12th Sanitary Engineering Conference on Nitrate and Water Supply: Source and Control.* Univ. of Illinois, Urbana, College of Engineering Publication, pp 14-26, 1970.

514. Stout, P. R., et al, *A Study of the Vertical Movement of Nitrogenous Matter from the Ground Surface to the Water Table in the Vicinity of Grover City and Arroyo Grande, San Luis Obispo County*, Research Report, Univ. of California, Davis, Dept. of Soils and Plant Nutrition, 51 pp, Jan. 1965.

515. California Dept. of Water Resources, *Delano Nitrate Investigation*, Bulletin 143-6, 42 pp, 1968.

516. Environment Staff Report, "Poisoning the Wells," *Environment,* Vol. 11, No. 1, pp 16-23, 45, 1969.

517. Nightingale, H. I., "Statistical Evaluation of Salinity and Nitrate Content and Trends Beneath Urban and Agricultural Area — Fresno, California," *Ground Water,* Vol. 8, No. 1, pp 22-28, 1970.

518. Schmidt, K. D., "The Use of Chemical Hydrographs in Ground-water Quality Studies," *Hydrology and Water Resources in Arizona and the Southwest,* Proceedings of Arizona Section-Amer. Water Resources Assoc. and the Hydrology Section-Arizona Academy of Science, Vol. 1, pp 211-223, 1971.

519. Schmidt, K. D., "Nitrate in Ground Water of the Fresno-Clovis Metropolitan Area, California," *Ground Water,* Vol. 10, No. 1, pp 50-61, 1972.

520. Nightingale, H. I., "Nitrates in Soil and Groundwater Beneath Irrigated and Fertilized Crops," *Soil Science,* Vol. 114, No. 4, pp 300-311, 1972.

521. Willardson, L. S., et al, "Drain Installation for Nitrate Reduction," *Ground Water,* Vol. 8, No. 4, pp 11-13, 1970.

522. Willardson, L. S., et al, "Nitrate Reduction with Submerged Drains," *Trans. Amer. Soc. of Agricultural Engineers,* Vol. 15, No. 1, pp 84-85, 90, 1972.

523. Pratt, P. F., et al, "Nitrate in Deep Soil Profiles in Relation to Fertilizer Rates and Leaching Volume," *Jour. of Environmental Quality,* Vol. 1, No. 1, 97-102, 1972.

524. Pratt, P. F., *Nitrate in the Unsaturated Zone Under Agricultural Lands,* U.S. Environmental Protection Agency Report EPA-16060-DOE-04/72, Water Pollution Control Research Series, 45 pp, April 1972.
(NTIS: PB-211 166)

525. Ayers, R. S., and R. L. Branson (editors), *Nitrates in the Upper Santa Ana River Basin in Relation to Groundwater Pollution,* Calif. Agric. Exp. Sta. Bull. 861, 59 pp, May 1973.

526. Vogt, J. E., "Infectious Hepatitis at Posen, Michigan," *Jour. Amer. Water Works Assoc.,* Vol. 53, No. 10, pp 1238-1242, 1961.

527. Hancock, J. C., "Public Health Aspects of Individual Water Wells," *Ground Water,* Vol. 1, No. 3, pp 27-29, 1963.

528. Foster, H. B., Jr., and C. L. Young, "Chlorination of Ground-water Supplies," *Jour. Amer. Water Works Assoc.,* Vol. 58, No. 11, pp 1471-1474, 1966.

529. Robeck, G., "Microbial Problems in Groundwater," *Ground Water,* Vol. 7, No. 3, pp 33-35, 1969.

530. Benson, W. W., et al, "An Apparent Case of Pesticide Poisoning," *Public Health Reports,* Dept. of Health, Education and Welfare, Vol. 85, pp 600-602, July 1970.

531. Hodges, A. L., Jr., et al, *Gas and Brackish Water in Fresh-Water Aquifers, Lake Charles Area, Louisiana,* Louisiana Geological Survey Water Resources Pamphlet 13, 35 pp, 1963.

532. Harder, A. H., et al, *Methane in the Fresh-Water Aquifers of Southwestern Louisiana and Theoretical Explosion Hazards,* Louisiana Geological Survey Water Resources Pamphlet 14, 22 pp, 1965.

533. U.S. Federal Water Pollution Control Admin., *Improved Sealants for Infiltration Control, The Development and Demonstration of Materials to Reduce or Eliminate Water Infiltration into Sewage,* Water Pollution Control Research Series, Report WP-20-18, 96 pp, June 1969.
(NTIS: PB-185 950)

534. Grossman, I. G., "Waterborne Styrene in a Crystalline Bedrock Aquifer in the Gales Ferry Area, Ledyard, Southeastern Connecticut," *Geological Survey Research 1970,* U.S. Geological Survey Prof. Paper 700-B, pp B203-B209, 1970.

535. Dixon, N., and D. W. Hendricks, "Simulation of Spatial and Temporal Changes in Water Quality Within a Hydrologic Unit," *Water Resources Bulletin,* Vol. 6, No. 4, pp 483-497, 1970.

536. Folkman, Y., and A. M. Wachs, "Filtration of Chlorella Through Dune-Sand," *Jour. Sanitary Engineering Div.,* Amer. Soc. of Civil Engineers, Vol. 96, No. SA3, pp 675-690, 1970.

537. Aley, T. J., et al, *Groundwater Contamination and Sinkhole Collapse Induced by Leaky Impoundments in Soluble Rock Terrain,* Engineering Geological Series No. 5, Missouri Geological Survey and Water Resources, Rolla, 32 pp, 1972.

538. Ferris, J. G., "Ground-Water Aquifers as Waste-Disposal Reservoirs — An Outline of the Basic Hydrologic Problems Involved," *Proceedings of 6th Annual Pollution Abatement Conf.,* Manfacturing Chemists Assoc., Inc., Washington, D. C., pp 68-74, 1951.

539. Theis, C. V., "Geologic and Hydrologic Factors in Ground Disposal of Waste," *1954 Sanitary Engineering Conf.,* U.S. Atomic Energy Commission Report WASH-275, pp 261-283, 1955.

540. LeGrand, H. E., "System for Evaluation of Contamination Potential of Some Waste Disposal Sites," *Jour. Amer. Water Works Assoc.,* Vol. 56, No. 8, pp 959-974, 1964.

541. DeBuchananne, G. D., and P. E. LaMoreaux, "Geologic Controls Related to Ground-Water Contamination," *Water Well Jour.,* Vol. 16, No. 3, pp 8, 40-44, 1962.

542. Anon., "Movements of Contaminants Through Geologic Formations," *Water Well Jour.,* Vol. 16, No. 3, pp 12-13, 1962.

543. Deutsch, M., "Natural Controls Involved in Shallow Aquifer Contamination," *Ground Water,* Vol. 3, No. 3, pp 37-40, 1965.

544. Pfannkuch, H. O., and P. K. Saint, "Hydrogeologic Framework for Deterioration in Groundwater Quality," *Proceedings of Conf. Toward a Statewide Ground Water Quality Information System,* Water Resources Research Center, Univ. of Minnesota, Minneapolis, pp 35-58, Feb. 1973.

545. Stewart, J. W., et al, *Geologic and Hydrologic Investigation at the Site of the Georgia Nuclear Laboratory, Dawson County, Georgia,* U.S. Geological Survey Bulletin 1133-F, pp F1-F90, 1964.

546. Maxey, G. B., and R. N. Farvolden, "Hydrogeologic Factors in Problems of Contamination in Arid Lands," *Ground Water,* Vol. 3, No. 4, pp 29-32, 1965.

547. Morris, D. A., "Use of Chemical and Radioactive Tracers at the National Reactor Testing Station, Idaho," *Isotope Techniques in the Hydrologic Cycle,* Geophysical Monograph Series, No. 11, Amer. Geophysical Union, pp 130-142, 1967.

548. Marine, I. W., "The Use of a Tracer Test to Verify an Estimate of the Groundwater Velocity in Fractured Crystalline Rock at the Savannah River Plant near Aiken, South Carolina," *Isotope Techniques in the Hydrologic Cycle,* Geophysical Monograph Series, No. 11, Amer. Geophysical Union, pp 171-179, 1967.

549. Webster, D. S., et al, *Two-Well Tracer Test in Fractured Crystalline Rock,* U.S. Geological Survey Water-Supply Paper 1544-I, 22 pp, 1970.

550. Armstrong, F. E., et al, *Tritiated Water as a Tracer in the Dump Leaching of Copper,* Bureau of Mines Report of Investigations RI 7510, 39 pp, May 1971.

551. Brown, R. H., "Hydrologic Factors Pertinent to Ground-Water Contamination," *Ground Water,* Vol. 2, No. 1, pp 5-12, 1964.

552. McGauhey, P. H., and R. B. Krone, *Soil Mantle as a Wastewater Treatment System,* SERL Report No. 67-11, Sanitary Engineering Research Lab., Univ. of California, Berkeley, 201 pp, 1967.

553. Henry, H. R., "The Effects of Temperature and Density Gradients Upon the Movement of Contaminants in Saturated Aquifers," *International Assoc. of Scientific Hydrology,* Publication No. 78, pp 54-65, 1968.

554. Boyd, J. W., et al, *Bacterial Response to the Soil Environment,* Sanitary Engineering Paper No. 5, Colorado State Univ., Fort Collins, 22 pp, June 1969. (NTIS: PB-188 515)

555. Champlin, J. B. F., *The Relation of Ion Movement to Fine Particle Displacement in a Sand Bed,* Water Resources Center WRC-0369, Georgia Inst. of Technology, Atlanta, 22 pp, July 1969. (NTIS: PB-187 521)

556. Champlin, J. B. F., "The Physics of Fine-Particle Movement Through Permeable Aquifers," *Soc. of Petroleum Engineers Jour.,* Vol. 11, No. 4, pp 367-373, 1971.

557. Hajek, B. F., "Chemical Interactions of Wastewater in a Soil Environment," *Jour. Water Pollution Control Fed.*, Vol. 41, No. 10, pp 1775-1786, Oct. 1969.

558. Romero, J. C., "The Movement of Bacteria and Viruses Through Porous Media," *Ground Water*, Vol. 8, No. 2, pp 37-48, 1970.

559. Butler, R. G., et al, "Underground Movement of Bacterial and Chemical Pollutants," *Jour. Amer. Water Works Assoc.*, Vol. 46, No. 2, pp 97-111, 1954.

560. Fournelle, H. J., et al, "Experimental Groundwater Pollution at Anchorage, Alaska," *Public Health Reports*, Vol. 72, No. 3, pp 203-209, 1957.

561. Sampayo, F. F., and H. R. Wilke, "Temperature and Phosphates as Groundwater Tracers," *Ground Water*, Vol. 1, No. 4, pp 36-38, 1963.

562. Page, H. G., et al, "Behavior of Detergents (ABS), Bacteria, and Dissolved Solids in Water-Saturated Soils," *Short Papers in Geology, Hydrology, and Topography*, U.S. Geological Survey Prof. Paper 450-E, pp E179-E181, 1963.

563. Young, R. H. F., et al, *Travel of ABS and Ammonia Nitrogen with Percolating Water Through Saturated Oahu Soils*, Technical Report 1, Water Resources Research Center, Univ. of Hawaii, Honolulu, 54 pp, Jan. 1967.

564. Kumagai, J. S., *Infiltration and Percolation Studies of Sulfides and Sewage Carbonaceous Matter*, Technical Report 7, Water Resources Research Center, Univ. of Hawaii, Honolulu, 58 pp, June 1967.

565. Ishizaki, K., et al, *Effects of Soluble Organics on Flow Through Thin Cracks of Basaltic Lava*, Technical Report 16, Water Resources Research Center, Univ. of Hawaii, Honolulu, 56 pp, Aug. 1967.

566. Krone, R. B., "The Movement of Disease Producing Organisms Through Soils," *Symposium on Municipal Sewage Effluent for Irrigation*, Louisiana Polytechnical Institute, Ruston, pp 75-104, 1968.

567. Crosby, J. W., III, et al, "Migration of Pollutants in a Glacial Outwash Environment," *Water Resources Research,* Vol. 4, No. 5, pp 1095-1114, 1968.

568. Crosby, J. W., III, et al, "Migration of Pollutants in a Glacial Outwash Environment, 3," *Water Resources Research,* Vol. 7, No. 3, pp 713-720, 1971.

569. Scalf, M. R., et al, "Movement of DDT and Nitrates During Ground-Water Recharge," *Water Resources Research,* Vol. 5, No. 5, pp 1041-1051, 1969.

570. Scalf, M. R., et al, *Fate of DDT and Nitrate in Ground Water,* U.S. Federal Water Pollution Control Admin., Robert S. Kerr Water Research Center, Ada, Okla., and U. S. Agricultural Research Service, Southwestern Great Plains Research Center, Bushland, Texas, 46 pp, April 1968.

571. Tilstra, J. R., et al, "Removal of Phosphorus and Nitrogen from Waste-Water Effluents by Induced Soil Percolation," *Jour. Water Pollution Control Fed.,* Vol. 44, No. 5, pp 796-805, 1972.

572. Wentink, G. R., and J. E. Etzel, "Removal of Metal Ions by Soil," *Jour. Water Pollution Control Fed.,* Vol. 44, No. 8, pp 1561-1574, 1972.

573. Allen, M. J., and S. M. Morrison, "Bacterial Movement Through Fractured Bedrock," *Ground Water,* Vol. 11, No. 2, pp 6-10, 1973.

574. California Dept. of Public Works, *Ground Water Quality Monitoring Program in California,* Water Quality Investigation Report No. 14, Sacramento, 198 pp, June 1956.

575. Bookman and Edmonston, Consulting Civil Engineers, *Activities of Public Agencies in Water Quality Investigations and Water Pollution Control in the San Gabriel River System,* Report to Central and West Basin Water Assoc., 28 pp, Oct. 1962.

576. Pomeroy, R. D., and G. T. Orlob, *Problems of Setting Standards and of Surveillance for Water Quality Control,* California State Water Quality Control Board, Publication No. 36, 123 pp, 1967.

577. Moreland, J. A., and J. A. Singer, *Evaluation of Water-Quality Monitoring in the Orange County Water District, California*, U.S. Geological Survey Open-File Report, 27 pp, 1969.

578. Water Resources Engineers, Inc., *An Investigation of Salt Balance in the Upper Santa Ana River Basin*, Final Report to the California State Water Resources Control Board and the Santa Ana River Basin Regional Water Quality Control Board, 2 Vols., 198 pp, 1969, 1970.

579. California State Water Resources Control Board, *Evaluation of Water Quality Monitoring Programs in California*, Sacramento, 57 pp, Feb. 1971.

580. Rainwater, F. H., and L. L. Thatcher, *Methods for Collection and Analysis of Water Samples*, U.S. Geological Survey Water-Supply Paper 1454, 301 pp, 1960.

581. Federal Interagency Work Group on Designation of Standards for Water Data Acquisition, *Recommended Methods for Water-Data Acquisition*, Preliminary Report, U.S. Geological Survey Office of Water Data Coordination, Washington, D. C., 412 pp, 1972.

582. Environmental Instrumentation Group, *Instrumentation for Environmental Monitoring—Water*, LBL-1, Vol. 2, Lawrence Berkeley Lab., Univ. of California, Berkeley, Feb. 1973.

583. Morgan, C. O., et al, "Digital Computer Methods for Water-Quality Data," *Ground Water*, Vol. 4, No. 3, pp 35-42, 1966.

584. Simpson, E. E., et al, *Space Time Sampling of Pollutants in Aquifers*, Symposium on Ground-Water Pollution, 15th General Assembly of International Union of Geodesy and Geophysics, Moscow, U.S.S.R., August 1971.

585. Cearlock, D. B., "A Systems Approach to Management of the Hanford Ground-Water Basin," *Ground Water*, Vol. 10, No. 1, pp 88-98, 1972.

586. Lee, E. S., *Analysis, Modeling and Forecasting of Stochastic Water Quality Systems*, Kansas Water Resources Inst., Contrib. No. 110, Kansas State Univ., Manhattan, 2 Vols., 498 pp, 1972.

587. Zaporozec, A., "Graphical Interpretation of Water-Quality Data," *Ground Water*, Vol. 10, No. 2, pp 32-43, 1972.

588. Moore, S. F., "Estimation Theory Applications to Design of Water Quality Monitoring Systems," *Jour. Hydraulics Div.,* Amer. Soc. of Civil Engineers, Vol. 99, No. HY5, pp 815-831, 1973.

589. McMillion, L. G., and J. W. Keeley, "Sampling Equipment for Ground-Water Investigations," *Ground Water,* Vol. 6, No. 2, pp 9-11, 1968.

590. LeGrand, H. E., "Monitoring of Changes in Quality of Ground Water," *Ground Water,* Vol. 6, No. 3, pp 14-18, 1968.

591. Warner, D. L., "Preliminary Field Studies Using Earth Resistivity Measurements for Delineating Zones of Contaminated Ground Water," *Ground Water,* Vol. 7, No. 1, pp 9-16, 1969.

592. Peterson, F. L., and C. Lao, "Electric Well Logging of Hawaiian Basaltic Aquifers," *Ground Water,* Vol. 8, No. 2, pp 11-18, 1970.

593. Turcan, A. N., Jr., and A. G. Winslow, "Quantitative Mapping of Salinity, Volume, and Yield of Saline Aquifers Using Borehole Geophysical Logs," *Water Resources Research,* Vol. 6, No. 5, pp 1478-1481, 1970.

594. Brown, D. L., "Techniques for Quality-of-Water Interpretations from Calibrated Geophysical Logs, Atlantic Coastal Area," *Ground Water,* Vol. 9, No. 4, pp 25-38, 1971.

595. Foster, J. B., and D. A. Goolsby, *Construction of Waste-Injection Monitor Wells Near Pensacola, Florida,* Florida Dept. of Natural Resources, Div. of Interior Resources, Bureau of Geology Information Circular No. 74, Tallahassee, 34 pp, 1972.

596. Andersen, J. R., "Groundwater Quality Studies at Waste Disposal Sites," *South Dakota's Environment, Its Pollution and Preservation — Symposium Proceedings,* South Dakota State Univ., pp 25-37, 1971.

597. Andersen, J. R., and J. N. Dornbush, "A Study of the Influence of a Sanitary Landfill on Ground Water Quality," *Annual Report for Fiscal Year 1966,* South Dakota State Univ., Water Resources Inst., pp 1-27, 1966.

598. Anon., *Ground Water Quality and Treatment,* Proceedings 14th Water Quality Conf., Univ. of Illinois, Urbana, 1973.

599. Appel, C. A., *Salt-Water Encroachment into Aquifers of the Raritan Formation in the Sayerville Area, Middlesex County, New Jersey, With a Section on a Proposed Tidal Dam on the South River,* Special Report No. 17, New Jersey Div. of Water Policy and Supply, 47 pp, 1962.

600. Becher, A. E., *Hydrogeologic Controls on and Water-Quality Effects of a Gasoline Spill Near Mechanicsburg, Pennsylvania,* U.S. Geological Survey Open-File Report, 6 pp, 1972.

601. Biggar, J. W., et al, *Soil Interaction with Organically Polluted Water,* Summary Report, Dept. of Water Science and Engineering, Univ. of California, Davis, Feb. 1966.

602. Bookman and Edmonston, Consulting Civil Engineers, *Management of Ground Water Quality in the Central and West Basin Water Replenishment District,* Report to the Central and West Basin Water Replenishment District, Downey, Calif., Nov. 1970.

603. Bryson, W. R., *The Occurrence of Salty Ground Water in the Shaffer Area, Rush County, Kansas,* Kansas Dept. of Health, Environmental Health Services, 14 pp, 1970.

604. Bryson, W. R., et al, *Residual Salt Study of Brine Affected Soil and Shale, Potwin Area, Butler County, Kansas,* Kansas Dept. of Health, Environmental Health Services Bulletin 3-1, 28 pp, 1966.

605. Burnitt, S. C., *Investigation of Ground-Water Contamination, City of Hawkins, Wood County, Texas,* Texas Water Comm. Report LD-0162-MR, 27 pp, 1963.

606. Burnitt, S. C., *Investigation of Ground-Water Contamination, Henderson Oil Field Area, Rusk County, Texas,* Texas Water Comm. Report LD-0262-MR, 14 pp, 1962.

607. Burnitt, S. C., and R. L. Crouch, *Investigation of Ground-Water Contamination, PHD, Hackberry, and Storie Oil Fields, Garza County, Texas,* Texas Water Comm. Report LD-0764, 94 pp, 1964.

608. Butcher, D. L., *The Occurrence of Salty Ground Water in the Albert Area, Barton County, Kansas,* Kansas Dept. of Health, Environmental Health Services, 19 pp, 1971.

609. Childs, K. E., *History of the Salt, Brine, and Paper Industries and Their Probable Effect on the Ground Water Quality in the Manistee Lake Area of Michigan*, Bureau of Water Management, Michigan Dept. of Natural Resources, 75 pp, 1970.

610. Committee on Geologic Aspects of Radioactive Waste Disposal, *Report to the U.S. Atomic Energy Commission*, National Academy of Sciences, Washington, D. C., 1966.

611. Cooper, W., *Possible Ground-Water Contamination in the Rolling Hills Addition, Potter County, Texas*, Texas Water Development Board, CL-6802, Aug. 1970.

612. Crouch, R. L., *Investigation of Alleged Ground-Water Contamination, Tri-Rue and Ride Oil Fields, Scurry County, Texas*, Texas Water Comm. Report LD-0464-MR, 18 pp, 1964.

613. Crouch, R. L., *Investigation of Ground-Water Contamination in the Juliana and West Jud Oil Fields, Haskell and Stonewall Counties, Texas*, Texas Water Comm. Report LD-0364-MR, 20 pp, 1964.

614. Crouch, R. L., and S. C. Burnitt, *Investigation of Ground-Water Contamination in the Vealmoor Oil Field, Howard and Borden Counties, Texas*, Texas Water Comm. Report LD-0265, 55 pp, 1965.

615. Dotson, G. K., et al, *Land Spreading, A Conserving and Non-Polluting Method of Disposing of Oily Wastes*, Adv. Waste Treatment Research Lab., Federal Water Quality Admin., Cincinnati, Ohio, 1970.

616. Draper, D. C., *Investigation of Contamination Complaint in South-Central Knox County, Texas*, Texas Board of Water Engineers Contamination Report No. 7, 8 pp, 1960.

617. Eliassen, R., et al, *Studies on the Movement of Viruses in Ground Water*, Annual Report, Commission on Environment Hygiene, Stanford Univ., 1965.

618. Emrich, G. H., and R. A. Landon, *Investigation of the Effects of Sanitary Landfills in Coal Strip Mines on Ground Water Quality*, Pennsylvania Bureau of Water Quality Management Publication No. 30, 37 pp, 1971.

619. Eto, M. A., et al, *Behavior of Selected Pesticides with Percolating Water in Oahu Soils,* Water Resources Research Center, Univ. of Hawaii, Honolulu, Aug. 1967.

620. Fink, B. E., *Investigation of Ground-Water Contamination by Cotton Seed Delinting Acid Waste, Terry County, Texas,* Texas Water Comm. Report LD-0864, 25 pp, 1964.

621. George, A., "Cave Pollution Can Mean Ground Water Pollution," *Ground Water Age,* Vol. 5, No. 6, pp 20-24, 1971.

622. Gold, D. P., et al, "Water Well Explosions — An Environmental Hazard," *Earth and Mineral Science,* Vol. 40, No. 3, pp 17-21, 1970.

623. Hanway, J. J., et al, *The Nitrate Problem,* Special Report No. 34, Iowa State Univ. Coop. Extension Service, 20 pp, 1963.

624. Holloway, H. D., *Bacteriological Pollution of Ground Water in the Big Spring Area, Howard County, Texas,* Texas Water Comm. Report LD-0163-MR, 14 pp, 1963.

625. Holloway, H. D., *Investigation of Ground-Water Contamination, City of Valera, Coleman County, Texas,* Texas Water Comm. Report LD-0362-MR, 7 pp, 1962.

626. Holloway, H. D., *Investigation of Alleged Ground-Water Contamination Near Kilgore, Gregg County, Texas,* Texas Water Comm. Report LD-0664, 15 pp, 1964.

627. Hutchinson, F. E., *The Influence of Salts Applied to Highways on the Levels of Sodium and Chloride Ions Present in Water and Soil Samples,* Technical Report, Water Resources Research Inst., Univ. of Maine, 20 pp, 1969.

628. Keech, D. K., "Ground Water Pollution," *Principles and Applications of Ground Water Hydraulics Conf.,* Michigan State Univ., East Lansing, 20 pp, 1970.

629. Kilpatrick, F. J., "Sanitation Problems in Unsewered Areas," *Minnesota Municipalities,* Vol. 44, pp 315-    , 1959.

630. Lane, J. W., and R. Newcome, Jr., *Status of Salt-Water Encroachment in Aquifers Along the Mississippi Gulf Coast, 1964,* Mississippi Board of Water Commissioners Bulletin 64-5, 16 pp, 1964.

631. Lehr, J. H., *A Study of the Ground Water Contamination Due to Saline Water Disposal in the Morrow County (Ohio) Oil Fields,* Ohio Water Resources Center, Ohio State Univ., 81 pp, March 1969.

632. Leonard, R. B., *Variations in Chemical Quality of Ground Water Beneath an Irrigated Field, Cedar Bluffs Irrigation District, Kansas,* Kansas Dept. of Health Bulletin 1-11, 20 pp, 1969.

633. Littleton, R. T., *Contamination of Surface and Ground Water in Southeast Young County, Texas,* Texas Board of Water Engineers, 13 pp, 1956.

634. Loehr, R. C., "Control of Nitrogen from Animal Waste Waters," *Proceedings 12th Sanitary Engineering Conf.,* Univ. of Illinois, Urbana, pp 177-186, Feb. 1970.

635. Loehr, R. C., *Pollution Implications of Animal Wastes – A Forward Oriented Review,* Robert S. Kerr Water Research Center, Federal Water Pollution Control Admin., 1968.

636. Love, J. D., and L. Hoover, *A Summary of the Geology of Sedimentary Basins of the United States with Reference to the Disposal of Radioactive Wastes,* U.S. Geological Survey Trace Elements Investigation Report 768, Open File, 92 pp, 1960.

637. Metzler, D. F., *An Investigation of the Sources and Seasonal Variations of Nitrates in Private and Public Water Supply Wells, Particularly with Respect to the Occurrence of Infant Cyanosis,* Final Report, Project No. RG4775, Univ. of Kansas, 33 pp, 1958.

638. Middleton, F. M., *Report on Analysis of Organic Contaminants Recovered from Town and Country Mutual Water Company Well at Commercetown, Colorado,* R. A. Taft Engineering Center, Cincinnati, Ohio, 1957.

639. Miller, R. A., and S. W. Maher, *Geologic Evaluation of Sanitary Landfill Sites in Tennessee,* Environmental Geology Series No. 1, Tennessee Dept. of Conservation, 38 pp, 1972.

640. Owens, W. G., *Protection of an Aquifer – A Case History,* Amer. Inst. of Mining, Metallurgy, and Petroleum Engineers, Paper No. SPE 3617, 19 pp, 1971.

641. Peckham, R. C., *Investigation of Contamination Complaint, Clemens Prison Farm, Brazoria County, Texas,* Texas Board of Water Engineers Contamination Report 9, 8 pp, 1960.

642. Pratt, P. F., *Quality Criteria for Trace Elements in Irrigation Waters,* Div. of Agricultural Sciences, Univ. of California, Riverside, 46 pp, 1972.

643. Shamburger, V. M., Jr., *Memorandum Report on Water Well Contamination in the Saspamco Area, Wilson County, Texas,* Texas Board of Water Engineers Contamination Report No. 3, 13 pp, 1958.

644. Shamburger, V. M., Jr., *A Reconnaissance of Alleged Salt-Contamination of Soils Near Stamford, Jones County, Texas,* Texas Board of Water Engineers Contamination Report No. 6, 8 pp, 1960.

645. Shamburger, V. M., Jr., *Reconnaissance of Alleged Water Well Contamination in the Garwood-Nada Area, Colorado County, Texas,* Texas Board of Water Engineers, 8 pp, 1959.

646. Shamburger, V. M., Jr., *Reconnaissance Report on Alleged Contamination of California Creek Near Avoca, Jones County, Texas,* Texas Board of Water Engineers Contamination Report No. 5, 14 pp, 1958.

647. Shamburger, V. M., Jr., *Reconnaissance of Water Well Pollution and the Occurrence of Shallow Ground Water, Runnels County, Texas,* Texas Board of Water Engineers Contamination Report No. 1, 38 pp, 1958.

648. Sherwood, C. B., and R. G. Grantham, *Water Control vs. Seawater Intrusion, Broward County, Florida,* Florida Geological Survey Leaflet 5, 17 pp, 1966.

649. Smith, W. W., "Salt Water Disposal: Sense and Dollars," *Petroleum Engineer,* Vol. 42, No. 11, pp 64-72, 1970.

650. Stead, F. W., "Groundwater Contamination," *Symposium on Education for the Peaceful Uses of Nuclear Explosives,* Tucson, Ariz., 1970.

651. Stearman, J., *A Reconnaissance Investigation of Alleged Contamination of Irrigation Wells Near Lockett, Wilbarger County, Texas,* Texas Board of Water Engineers Contamination Report No. 8, 12 pp, 1960.

652. Thornhill, J. T., *Investigation of Ground-Water Contamination, Coleto Creek Oil Field, Victoria County, Texas,* Texas Water Comm. Report LD-0564-MR, 21 pp, 1964.

653. U.S. Army Corps of Engineers, *Report on Ground Water Contamination, Rocky Mountain Arsenal, Denver, Colorado,* 31 pp, 1955.

654. U.S. Federal Water Pollution Control Admin., *A Report on the Examination of the Waste Treatment and Disposal Operations at the National Reactor Testing Station, Idaho Falls, Idaho,* Northwest Region (Now EPA, Region X, Seattle), 1970.

655. U.S. Public Health Service, *Ground-Water Pollution in the South Platte River Valley Between Denver and Brighton, Colorado,* South Platte River Basin Project PR-4, 1965.

656. U.S. Public Health Service, *Water Well Contamination and Waste Disposal in the Greater Anchorage Area,* U.S. Public Health Service Report, 3 pp, 1965.

657. Walters, K. L., *Reconnaissance of Sea-Water Intrusion Along Coastal Washington, 1966-68,* Washington Dept. of Ecology Water-Supply Bulletin 32, 208 pp, 1971.

658. Warner, D. L., *Survey of Industrial Waste Injection Wells,* Final Report, U.S. Geological Survey Contract No. 14-08-0010-12280, Univ. of Missouri, Rolla, 3 Vols., 1972.

659. Wells, D., et al, *Potential Pollution of the Ogallala by Recharging Playa Lake Water — Pesticides,* U.S. Environmental Protection Agency, Water Pollution Control Research Series Report EPA-16060-DCO-10/70, Oct. 1970.
(NTIS: PB-208 813)

660. White, J. W., *Investigation of Salt Water Contamination in a Woodbine Well Near Sherman, Grayson County, Texas,* Texas Board of Water Engineers Contamination Report 10, 10 pp, 1961.

661. Wilson, C. B., and T. H. Essig (Editors), *Evaluation of Radiological Conditions in the Vicinity of Hanford for 1969*, BNWL-1505, Battelle Labs., Richland, Wash., 1970.

# AUTHOR INDEX

Numbers refer to References.